Choosing the Perfect Spring for Your Barreled Spring Clock

By Richard Hansen

Goofy Rooster Publishing
Wylie, Texas
www.goofyrooster-publishing.com

Choosing the Perfect Spring for Your Barreled Spring Clock

By Richard Hansen

Goofy Rooster Publishing   Wylie, Texas 75098
www.goofyrooster-publishing.com

ISBN-10: 0-9897136-0-1

ISBN-13: 978-0-9897136-0-3

Dedicated to my mother,

who was successful in pushing me to become an electrical engineer.

Table of Contents

Purpose of This Book:

As I journeyed into the world of clock repair and restoration, I became acutely aware of a very common problem that seldom has an easy solution: if a clock on my workbench has missing springs (or has the wrong springs installed), how can I correctly pick a proper replacement main spring for it?

When customers bring me their clocks to be repaired, I (too) often see evidence of poor repairs, poor techniques and even gross mistakes inflicted on them by previous repairmen. Why then should I believe that every spring I find is, in fact, the perfect one for the clock in question? And when I find a spring where the outer end has obviously been re-worked (thus making the spring a bit shorter), how do I know if it's not now *too* short? Is it possible to redo the end again, or will that make the spring *impossibly* too short then?

This book is a result of my work to answer some of these questions about choosing springs. It provides a methodology whereby answers to those questions can be found by using a set of look-up tables. It's a solution intended for the use of clock repairmen at their bench to find those answers with a minimum of time and effort needed for any given clock in front of them.

I've written this book with the encouragement of two fellow clock repairmen: my brother Fred, and my good friend Gerald Grunsel.

- Richard Hansen

How to Use This Book:

To determine the correct spring for any clock, a repairman would need to "do the math"...if he knows the correct formulas. In this book, I've saved you the work and bother by doing most of the math for you (using the correct formulas).

When determining the best spring for any clock where the spring is enclosed in a barrel, the repairman has to take all three dimensions of the spring into consideration: the width, the thickness, and the length.

Spring Width:

The width is easy to find. It only requires one measurement, the barrel depth.

Measure the barrel inside depth: this is the dimension from the lip that the barrel lid sits on, down to the inside bottom of the barrel. Subtracting about 0.050 inches from the barrel depth will give you the maximum width of the spring. The 0.050 is to give you side clearance for the spring (0.025 on each side). Be sure that you have enough side clearance, because if the spring is too wide it will tend to push the cap off of the barrel.

There is no minimum width other than one half the barrel depth. If it's narrower than that, the spring will curl on itself. You want to keep the spring as wide as possible because its strength is proportional to its width and you always want the spring to be as strong as possible.

Spring Length:

You need to calculate how long the spring must be to run the clock for 168 turns of the minute hand (168 hours is the number of hours in one week).

You must measure:

- # of teeth on barrel
- # of leaves on pinion driven by barrel
- # of teeth on wheel that the pinion drives
- # of leaves on pinion that the wheel drives

For example:

Barrel teeth = 70
Leaves on pinion driven by barrel = 9
Teeth on wheel that the pinion drives = 63
Leaves on pinion that the wheel drives = 10

Turns of minute hand per one barrel turn = (70 times 63) divided by (9 times 10) = 49
Number of barrel turns per day = 49 divided by 24 = 2.04
Number of barrel turns per week = 2.04 times 7 = 14.28

So theoretical turns of 14.28 will drive the clock for 7 days...right? No, actually, you never get theoretical drive out of a spring...you only get about 70% of it. In our example, you would look for spring thickness in your barrel and arbor that would turn the barrel for 14.28 divided by 0.70 = 20.4 barrel turns for a week.

So, you need a total of 20.4 barrel turns to make up for the 30% you typically lose.

Spring Thickness:

Knowing the barrel inside diameter, arbor diameter and spring length, you can go to Chart A and determine the spring thickness that you want. Pick the thickest value you can to get the most power for the spring length.

Thickness is determined by the power required to drive the clock for at least 7 or 8 days. See Chart A for examples of spring thickness. The thicker the spring the shorter it must be to fit in the barrel, and the fewer turns you will be able to fit in the barrel.

Tools Needed

* Digital caliper (see photo below)

Use to measure spring thickness (from 0.008 thru 0.025 inches), and spring width (from 0.7 inches and up).

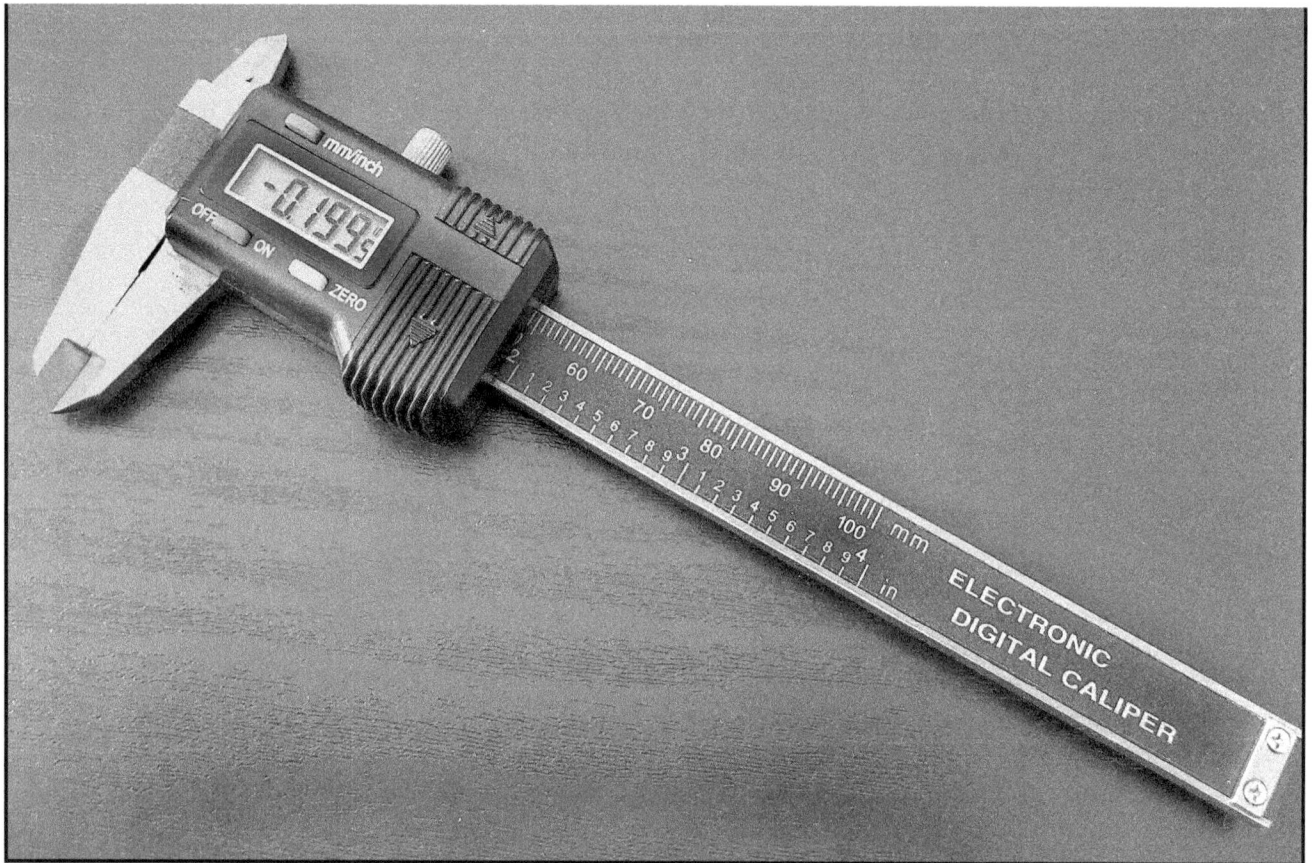

* Calculator

* Tape measure to measure spring lengths

* "C" clamp vice grips (see 2 photos on opposite page)

"C" clamp vice grips with jaws filed out for putting barrel caps back on. You'll need two of these, one to clamp one side of barrel cap, and the other to force the other side of cap on. Top photo on next page is of the basic "C" clamp vice grips; bottom photo on next page shows jaws filed out to accommodate pushing grip in far enough to not interfere with barrel center arbor. These allow better grip in the caps and avoid collapsing the caps with bent depressions.

Odd Notes

You should discard a spring:

- If it needs steel wool to clean it, and/or if it's rusted. A spring should be perfectly smooth, with no scratches as from steel wool.

- If it has excessive "flat" spots. You can feel these when you hand clean the spring...or you may be able to see bad wear patterns, or repeating rub marks on the spring.

- If it has any cracks.

A spring is a spring is a spring. Whether it is a spring in a Seth Thomas or in a 400 day clock, the same rules apply. Some parts catalogs note springs as just for 400 day clocks, but that is typically not true. The only exception is fuzee springs which may have a different strength versus winding characteristic.

Lubrication. I would never use anything on a spring which is called "grease", which implies thick viscosity. For the next guy's sake I never use graphite which is so messy. For best results, I use Mobil One synthetic oil.

A thinner spring will give you more turns in the barrel...but the strength will suffer.

Chart A shows that the actual length of a spring is not critical as to how many turns it will give you. So taking 3 or 4 inches off to make a new outer end, or taking 3 or 4 inches off a second time is not critical.

When the "Set" is bad.

"SET" is how coiled the spring is when it is unrestrained. What is the diameter of the spring when it is lying loose with no restraints or clamp on a table? Measure the diameter on a used spring just taken out of the clock BEFORE the spring is cleaned or stretched. A used spring may measure 4 or 5 inches, when a new one may measure 12 to 15 inches. See page 74 of this book for examples. The "set" of a spring should be at least 2 1/2 times the inside diameter of the barrel.

The area of a spring inside a barrel is the spring length multiplied by the spring thickness, which is the side view of the spring.

This book will not specifically discuss oiling, cleaning or winding of a spring. There is much literature done about those topics. Such a good article is "Everything You Always Wanted to Know About Clock Mainsprings" by Michael P. Murray, which can be found in the February 2004 issue of the American Watchmakers Institute.

All dimensions in this book are inches, unless otherwise noted.

Chart A: Picking a Spring Length

Given the barrel diameter, arbor diameter and spring thickness, the following charts allow you to pick the correct spring length. Each chart page is for a SPECIFIC barrel inside diameter.

If the Barrel Inside Dimension (in inches) is:	Use Chart	On Page
1.00	A-1.00	16-17
1.05	A-1.05	18-19
1.10	A-1.10	20-21
1.15	A-1.15	22-23
1.20	A-1.20	24-25
1.25	A-1.25	26-27
1.30	A-1.30	28-29
1.35	A-1.35	30-31
1.40	A-1.40	32-33
1.45	A-1.45	34-35
1.50	A-1.50	36-37
1.55	A-1.55	38-39
1.60	A-1.60	40-41
1.65	A-1.65	42-43
1.70	A-1.70	44-45
1.75	A-1.75	46-47
1.80	A-1.80	48-49
1.85	A-1.85	50-51
1.90	A-1.90	52-53
1.95	A-1.95	54-55
2.00	A-2.00	56-57
2.10	A-2.10	58-59
2.20	A-2.20	60-61
2.40	A-2.40	62-63

Understand that you never get best turns from a spring, partly because the inner spring end never fully uncoils.

The "best length" is true for the combination of barrel inside diameter, arbor diameter, and spring thickness. You cannot vary barrel inside diameter or arbor diameter, because they are fixed by your clock. You can vary spring thickness to get the length that you want.

Chart A-1.00: Only For Barrels With An Inside Diameter Of 1.00 Inches

(Arbor diameters of from 0.25 Inches thru 0.36 Inches, and spring thicknesses of from 0.009 Thru 0.016 Inches)

For An Arbor Diameter Of	And A Spring Thickness Of	The BEST Spring Length Is	Max Turns Using BEST Length	Max Turns If Spring Is 10" Short Or Long	For An Arbor Diameter Of	And A Spring Thickness Of	The BEST Spring Length Is	Max Turns Using BEST Length	Max Turns If Spring Is 10" Short Or Long
0.25	0.009	40.9	11.5	11.1	0.31	0.009	39.4	9.5	9.0
0.25	0.010	36.8	10.4	9.9	0.31	0.010	35.5	8.5	8.0
0.25	0.011	33.5	9.4	8.9	0.31	0.011	32.3	7.8	7.2
0.25	0.012	30.7	8.7	8.0	0.31	0.012	29.6	7.1	6.5
0.25	0.013	28.3	8.0	7.3	0.31	0.013	27.3	6.6	5.9
0.25	0.014	26.3	7.4	6.7	0.31	0.014	25.4	6.1	5.4
0.25	0.015	24.5	6.9	6.1	0.31	0.015	23.7	5.7	4.9
0.25	0.016	23.0	6.5	5.6	0.31	0.016	22.2	5.3	4.5
0.26	0.009	40.7	11.2	10.7	0.32	0.009	39.2	9.2	8.7
0.26	0.010	36.6	10.1	9.5	0.32	0.010	35.2	8.2	7.7
0.26	0.011	33.3	9.1	8.6	0.32	0.011	32.0	7.5	6.9
0.26	0.012	30.5	8.4	7.7	0.32	0.012	29.4	6.9	6.3
0.26	0.013	28.2	7.7	7.0	0.32	0.013	27.1	6.3	5.7
0.26	0.014	26.2	7.2	6.4	0.32	0.014	25.2	5.9	5.2
0.26	0.015	24.4	6.7	5.9	0.32	0.015	23.5	5.5	4.7
0.26	0.016	22.9	6.3	5.4	0.32	0.016	22.0	5.2	4.3
0.27	0.009	40.5	10.8	10.4	0.33	0.009	38.9	8.8	8.4
0.27	0.010	36.4	9.7	9.2	0.33	0.010	35.0	8.0	7.5
0.27	0.011	33.1	8.9	8.3	0.33	0.011	31.8	7.2	6.7
0.27	0.012	30.3	8.1	7.5	0.33	0.012	29.2	6.6	6.0
0.27	0.013	28.0	7.5	6.8	0.33	0.013	26.9	6.1	5.5
0.27	0.014	26.0	7.0	6.2	0.33	0.014	25.0	5.7	5.0
0.27	0.015	24.3	6.5	5.7	0.33	0.015	23.3	5.3	4.5
0.27	0.016	22.8	6.1	5.2	0.33	0.016	21.9	5.0	4.2
0.28	0.009	40.2	10.5	10.0	0.34	0.009	38.6	8.5	8.1
0.28	0.010	36.2	9.4	8.9	0.34	0.010	34.7	7.7	7.2
0.28	0.011	32.9	8.6	8.0	0.34	0.011	31.6	7.0	6.4
0.28	0.012	30.2	7.9	7.2	0.34	0.012	28.9	6.4	5.8
0.28	0.013	27.8	7.3	6.6	0.34	0.013	26.7	5.9	5.3
0.28	0.014	25.9	6.7	6.0	0.34	0.014	24.8	5.5	4.8
0.28	0.015	24.1	6.3	5.5	0.34	0.015	23.2	5.1	4.4
0.28	0.016	22.6	5.9	5.0	0.34	0.016	21.7	4.8	4.0
0.29	0.009	40.0	10.1	9.7	0.35	0.009	38.3	8.2	7.8
0.29	0.010	36.0	9.1	8.6	0.35	0.010	34.5	7.4	6.9
0.29	0.011	32.7	8.3	7.7	0.35	0.011	31.3	6.7	6.2
0.29	0.012	30.0	7.6	7.0	0.35	0.012	28.7	6.2	5.6
0.29	0.013	27.7	7.0	6.3	0.35	0.013	26.5	5.7	5.1
0.29	0.014	25.7	6.5	5.8	0.35	0.014	24.6	5.3	4.6
0.29	0.015	24.0	6.1	5.3	0.35	0.015	23.0	4.9	4.2
0.29	0.016	22.5	5.7	4.9	0.35	0.016	21.5	4.6	3.8
0.30	0.009	39.7	9.8	9.3	0.36	0.009	38.0	7.9	7.5
0.30	0.010	35.7	8.8	8.3	0.36	0.010	34.2	7.2	6.7
0.30	0.011	32.5	8.0	7.5	0.36	0.011	31.1	6.5	6.0
0.30	0.012	29.8	7.4	6.7	0.36	0.012	28.5	6.0	5.4
0.30	0.013	27.5	6.8	6.1	0.36	0.013	26.3	5.5	4.9
0.30	0.014	25.5	6.3	5.6	0.36	0.014	24.4	5.1	4.4
0.30	0.015	23.8	5.9	5.1	0.36	0.015	22.8	4.8	4.0
0.30	0.016	22.3	5.5	4.7	0.36	0.016	21.4	4.5	3.7

Richard Hansen

Chart A-1.00 (Continued): Only For Barrels With An Inside Diameter Of 1.00 Inches

(Arbor diameters of from 0.37 Inches thru 0.48 Inches, and spring thicknesses of from 0.009 Thru 0.016 Inches)

For An Arbor Diameter Of	And A Spring Thickness Of	The BEST Spring Length Is	Max Turns Using BEST Length	Max Turns If Spring Is 10" Short Or Long	For An Arbor Diameter Of	And A Spring Thickness Of	The BEST Spring Length Is	Max Turns Using BEST Length	Max Turns If Spring Is 10" Short Or Long
0.37	0.009	37.7	7.7	7.2	0.43	0.009	35.6	6.1	5.7
0.37	0.010	33.9	6.9	6.4	0.43	0.010	32.0	5.5	5.0
0.37	0.011	30.8	6.3	5.7	0.43	0.011	29.1	5.0	4.5
0.37	0.012	28.2	5.7	5.2	0.43	0.012	26.7	4.6	4.0
0.37	0.013	26.1	5.3	4.7	0.43	0.013	24.6	4.2	3.6
0.37	0.014	24.2	4.9	4.2	0.43	0.014	22.9	3.9	3.3
0.37	0.015	22.6	4.6	3.9	0.43	0.015	21.3	3.6	3.0
0.37	0.016	21.2	4.3	3.5	0.43	0.016	20.0	3.4	2.7
0.38	0.009	37.3	7.4	7.0	0.44	0.009	35.2	5.8	5.4
0.38	0.010	33.6	6.6	6.2	0.44	0.010	31.7	5.3	4.8
0.38	0.011	30.5	6.0	5.5	0.44	0.011	28.8	4.8	4.3
0.38	0.012	28.0	5.5	5.0	0.44	0.012	26.4	4.4	3.8
0.38	0.013	25.8	5.1	4.5	0.44	0.013	24.4	4.0	3.5
0.38	0.014	24.0	4.7	4.1	0.44	0.014	22.6	3.8	3.1
0.38	0.015	22.4	4.4	3.7	0.44	0.015	21.1	3.5	2.8
0.38	0.016	21.0	4.2	3.4	0.44	0.016	19.8	3.3	2.6
0.39	0.009	37.0	7.1	6.7	0.45	0.009	34.8	5.6	5.2
0.39	0.010	33.3	6.4	5.9	0.45	0.010	31.3	5.0	4.6
0.39	0.011	30.3	5.8	5.3	0.45	0.011	28.5	4.6	4.1
0.39	0.012	27.7	5.3	4.8	0.45	0.012	26.1	4.2	3.7
0.39	0.013	25.6	4.9	4.3	0.45	0.013	24.1	3.9	3.3
0.39	0.014	23.8	4.6	3.9	0.45	0.014	22.4	3.6	3.0
0.39	0.015	22.2	4.3	3.5	0.45	0.015	20.9	3.4	2.7
0.39	0.016	20.8	4.0	3.2	0.45	0.016	19.6	3.2	2.4
0.40	0.009	36.7	6.8	6.4	0.46	0.009	34.4	5.4	5.0
0.40	0.010	33.0	6.2	5.7	0.46	0.010	31.0	4.8	4.4
0.40	0.011	30.0	5.6	5.1	0.46	0.011	28.1	4.4	3.9
0.40	0.012	27.5	5.1	4.6	0.46	0.012	25.8	4.0	3.5
0.40	0.013	25.4	4.7	4.1	0.46	0.013	23.8	3.7	3.1
0.40	0.014	23.6	4.4	3.7	0.46	0.014	22.1	3.5	2.8
0.40	0.015	22.0	4.1	3.4	0.46	0.015	20.6	3.2	2.6
0.40	0.016	20.6	3.8	3.1	0.46	0.016	19.4	3.0	2.3
0.41	0.009	36.3	6.6	6.2	0.47	0.009	34.0	5.1	4.8
0.41	0.010	32.7	5.9	5.5	0.47	0.010	30.6	4.6	4.2
0.41	0.011	29.7	5.4	4.9	0.47	0.011	27.8	4.2	3.7
0.41	0.012	27.2	4.9	4.4	0.47	0.012	25.5	3.9	3.3
0.41	0.013	25.1	4.6	4.0	0.47	0.013	23.5	3.6	3.0
0.41	0.014	23.3	4.2	3.6	0.47	0.014	21.9	3.3	2.7
0.41	0.015	21.8	3.9	3.2	0.47	0.015	20.4	3.1	2.4
0.41	0.016	20.4	3.7	2.9	0.47	0.016	19.1	2.9	2.2
0.42	0.009	35.9	6.3	5.9	0.48	0.009	33.6	4.9	4.5
0.42	0.010	32.3	5.7	5.2	0.48	0.010	30.2	4.4	4.0
0.42	0.011	29.4	5.2	4.7	0.48	0.011	27.5	4.0	3.6
0.42	0.012	27.0	4.7	4.2	0.48	0.012	25.2	3.7	3.2
0.42	0.013	24.9	4.4	3.8	0.48	0.013	23.2	3.4	2.9
0.42	0.014	23.1	4.1	3.4	0.48	0.014	21.6	3.2	2.6
0.42	0.015	21.6	3.8	3.1	0.48	0.015	20.1	3.0	2.3
0.42	0.016	20.2	3.6	2.8	0.48	0.016	18.9	2.8	2.1

Selecting the Perfect Spring for Your Barreled Spring Clock

Chart A-1.05: Only For Barrels With An Inside Diameter Of 1.05 Inches

(Arbor diameters of from 0.25 Inches thru 0.36 Inches, and spring thicknesses of from 0.009 Thru 0.016 Inches)

For An Arbor Diameter Of	And A Spring Thickness Of	The BEST Spring Length Is	Max Turns Using BEST Length	Max Turns If Spring Is 10" Short Or Long	For An Arbor Diameter Of	And A Spring Thickness Of	The BEST Spring Length Is	Max Turns Using BEST Length	Max Turns If Spring Is 10" Short Or Long
0.25	0.009	45.4	12.6	12.2	0.31	0.009	43.9	10.5	10.1
0.25	0.010	40.8	11.3	10.9	0.31	0.010	39.5	9.4	9.0
0.25	0.011	37.1	10.3	9.8	0.31	0.011	35.9	8.6	8.1
0.25	0.012	34.0	9.4	8.9	0.31	0.012	32.9	7.8	7.3
0.25	0.013	31.4	8.7	8.1	0.31	0.013	30.4	7.2	6.7
0.25	0.014	29.2	8.1	7.4	0.31	0.014	28.2	6.7	6.1
0.25	0.015	27.2	7.5	6.8	0.31	0.015	26.3	6.3	5.6
0.25	0.016	25.5	7.1	6.3	0.31	0.016	24.7	5.9	5.2
0.26	0.009	45.2	12.2	11.8	0.32	0.009	43.6	10.1	9.7
0.26	0.010	40.6	11.0	10.5	0.32	0.010	39.3	9.1	8.7
0.26	0.011	36.9	10.0	9.5	0.32	0.011	35.7	8.3	7.8
0.26	0.012	33.9	9.2	8.6	0.32	0.012	32.7	7.6	7.1
0.26	0.013	31.3	8.5	7.8	0.32	0.013	30.2	7.0	6.4
0.26	0.014	29.0	7.8	7.2	0.32	0.014	28.1	6.5	5.9
0.26	0.015	27.1	7.3	6.6	0.32	0.015	26.2	6.1	5.4
0.26	0.016	25.4	6.9	6.1	0.32	0.016	24.5	5.7	5.0
0.27	0.009	44.9	11.8	11.4	0.33	0.009	43.4	9.8	9.4
0.27	0.010	40.4	10.7	10.2	0.33	0.010	39.0	8.8	8.4
0.27	0.011	36.8	9.7	9.2	0.33	0.011	35.5	8.0	7.5
0.27	0.012	33.7	8.9	8.3	0.33	0.012	32.5	7.4	6.8
0.27	0.013	31.1	8.2	7.6	0.33	0.013	30.0	6.8	6.2
0.27	0.014	28.9	7.6	7.0	0.33	0.014	27.9	6.3	5.7
0.27	0.015	27.0	7.1	6.4	0.33	0.015	26.0	5.9	5.2
0.27	0.016	25.3	6.7	5.9	0.33	0.016	24.4	5.5	4.8
0.28	0.009	44.7	11.5	11.1	0.34	0.009	43.1	9.5	9.1
0.28	0.010	40.2	10.3	9.9	0.34	0.010	38.8	8.5	8.1
0.28	0.011	36.6	9.4	8.9	0.34	0.011	35.2	7.8	7.3
0.28	0.012	33.5	8.6	8.1	0.34	0.012	32.3	7.1	6.6
0.28	0.013	30.9	8.0	7.4	0.34	0.013	29.8	6.6	6.0
0.28	0.014	28.7	7.4	6.7	0.34	0.014	27.7	6.1	5.5
0.28	0.015	26.8	6.9	6.2	0.34	0.015	25.8	5.7	5.0
0.28	0.016	25.1	6.5	5.7	0.34	0.016	24.2	5.3	4.6
0.29	0.009	44.4	11.1	10.7	0.35	0.009	42.8	9.2	8.8
0.29	0.010	40.0	10.0	9.6	0.35	0.010	38.5	8.3	7.8
0.29	0.011	36.4	9.1	8.6	0.35	0.011	35.0	7.5	7.0
0.29	0.012	33.3	8.4	7.8	0.35	0.012	32.1	6.9	6.4
0.29	0.013	30.8	7.7	7.1	0.35	0.013	29.6	6.4	5.8
0.29	0.014	28.6	7.2	6.5	0.35	0.014	27.5	5.9	5.3
0.29	0.015	26.7	6.7	6.0	0.35	0.015	25.7	5.5	4.9
0.29	0.016	25.0	6.3	5.5	0.35	0.016	24.1	5.2	4.5
0.30	0.009	44.2	10.8	10.4	0.36	0.009	42.5	8.9	8.5
0.30	0.010	39.8	9.7	9.3	0.36	0.010	38.2	8.0	7.6
0.30	0.011	36.1	8.8	8.3	0.36	0.011	34.7	7.3	6.8
0.30	0.012	33.1	8.1	7.6	0.36	0.012	31.8	6.7	6.1
0.30	0.013	30.6	7.5	6.9	0.36	0.013	29.4	6.1	5.6
0.30	0.014	28.4	6.9	6.3	0.36	0.014	27.3	5.7	5.1
0.30	0.015	26.5	6.5	5.8	0.36	0.015	25.5	5.3	4.7
0.30	0.016	24.9	6.1	5.3	0.36	0.016	23.9	5.0	4.3

Chart A-1.05 (Continued): Only For Barrels With An Inside Diameter Of 1.05 Inches

(Arbor diameters of from 0.37 Inches thru 0.48 Inches, and spring thicknesses of from 0.009 Thru 0.016 Inches)

For An Arbor Diameter Of	And A Spring Thickness Of	The BEST Spring Length Is	Max Turns Using BEST Length	Max Turns If Spring Is 10" Short Or Long	For An Arbor Diameter Of	And A Spring Thickness Of	The BEST Spring Length Is	Max Turns Using BEST Length	Max Turns If Spring Is 10" Short Or Long
0.37	0.009	42.1	8.6	8.2	0.43	0.009	40.0	6.9	6.6
0.37	0.010	37.9	7.7	7.3	0.43	0.010	36.0	6.2	5.8
0.37	0.011	34.5	7.0	6.6	0.43	0.011	32.8	5.7	5.2
0.37	0.012	31.6	6.4	5.9	0.43	0.012	30.0	5.2	4.7
0.37	0.013	29.2	5.9	5.4	0.43	0.013	27.7	4.8	4.3
0.37	0.014	27.1	5.5	4.9	0.43	0.014	25.7	4.5	3.9
0.37	0.015	25.3	5.1	4.5	0.43	0.015	24.0	4.2	3.5
0.37	0.016	23.7	4.8	4.1	0.43	0.016	22.5	3.9	3.2
0.38	0.009	41.8	8.3	7.9	0.44	0.009	39.7	6.7	6.3
0.38	0.010	37.6	7.5	7.0	0.44	0.010	35.7	6.0	5.6
0.38	0.011	34.2	6.8	6.3	0.44	0.011	32.4	5.5	5.0
0.38	0.012	31.4	6.2	5.7	0.44	0.012	29.7	5.0	4.5
0.38	0.013	28.9	5.7	5.2	0.44	0.013	27.5	4.6	4.1
0.38	0.014	26.9	5.3	4.7	0.44	0.014	25.5	4.3	3.7
0.38	0.015	25.1	5.0	4.3	0.44	0.015	23.8	4.0	3.4
0.38	0.016	23.5	4.7	4.0	0.44	0.016	22.3	3.8	3.1
0.39	0.009	41.5	8.0	7.6	0.45	0.009	39.3	6.4	6.1
0.39	0.010	37.3	7.2	6.8	0.45	0.010	35.3	5.8	5.4
0.39	0.011	33.9	6.5	6.1	0.45	0.011	32.1	5.3	4.8
0.39	0.012	31.1	6.0	5.5	0.45	0.012	29.5	4.8	4.3
0.39	0.013	28.7	5.5	5.0	0.45	0.013	27.2	4.4	3.9
0.39	0.014	26.7	5.1	4.6	0.45	0.014	25.2	4.1	3.6
0.39	0.015	24.9	4.8	4.2	0.45	0.015	23.6	3.9	3.3
0.39	0.016	23.3	4.5	3.8	0.45	0.016	22.1	3.6	3.0
0.40	0.009	41.1	7.7	7.4	0.46	0.009	38.9	6.2	5.8
0.40	0.010	37.0	7.0	6.5	0.46	0.010	35.0	5.6	5.2
0.40	0.011	33.6	6.3	5.9	0.46	0.011	31.8	5.1	4.6
0.40	0.012	30.8	5.8	5.3	0.46	0.012	29.2	4.6	4.2
0.40	0.013	28.5	5.3	4.8	0.46	0.013	26.9	4.3	3.8
0.40	0.014	26.4	5.0	4.4	0.46	0.014	25.0	4.0	3.4
0.40	0.015	24.7	4.6	4.0	0.46	0.015	23.3	3.7	3.1
0.40	0.016	23.1	4.3	3.7	0.46	0.016	21.9	3.5	2.8
0.41	0.009	40.8	7.5	7.1	0.47	0.009	38.5	5.9	5.6
0.41	0.010	36.7	6.7	6.3	0.47	0.010	34.6	5.3	5.0
0.41	0.011	33.4	6.1	5.6	0.47	0.011	31.5	4.9	4.4
0.41	0.012	30.6	5.6	5.1	0.47	0.012	28.9	4.5	4.0
0.41	0.013	28.2	5.2	4.6	0.47	0.013	26.6	4.1	3.6
0.41	0.014	26.2	4.8	4.2	0.47	0.014	24.7	3.8	3.3
0.41	0.015	24.5	4.5	3.9	0.47	0.015	23.1	3.6	3.0
0.41	0.016	22.9	4.2	3.5	0.47	0.016	21.6	3.3	2.7
0.42	0.009	40.4	7.2	6.8	0.48	0.009	38.1	5.7	5.4
0.42	0.010	36.4	6.5	6.1	0.48	0.010	34.2	5.1	4.8
0.42	0.011	33.1	5.9	5.4	0.48	0.011	31.1	4.7	4.3
0.42	0.012	30.3	5.4	4.9	0.48	0.012	28.5	4.3	3.8
0.42	0.013	28.0	5.0	4.4	0.48	0.013	26.3	4.0	3.5
0.42	0.014	26.0	4.6	4.0	0.48	0.014	24.5	3.7	3.1
0.42	0.015	24.2	4.3	3.7	0.48	0.015	22.8	3.4	2.9
0.42	0.016	22.7	4.0	3.4	0.48	0.016	21.4	3.2	2.6

Chart A-1.10: Only For Barrels With An Inside Diameter Of 1.10 Inches

(Arbor diameters of from 0.25 Inches thru 0.36 Inches, and spring thicknesses of from 0.009 Thru 0.016 Inches)

For An Arbor Diameter Of	And A Spring Thickness Of	The BEST Spring Length Is	Max Turns Using BEST Length	Max Turns If Spring Is 10" Short Or Long	For An Arbor Diameter Of	And A Spring Thickness Of	The BEST Spring Length Is	Max Turns Using BEST Length	Max Turns If Spring Is 10" Short Or Long
0.25	0.009	50.1	13.6	13.3	0.31	0.009	48.6	11.5	11.1
0.25	0.010	45.1	12.3	11.9	0.31	0.010	43.7	10.3	9.9
0.25	0.011	41.0	11.2	10.7	0.31	0.011	39.8	9.4	8.9
0.25	0.012	37.6	10.2	9.7	0.31	0.012	36.5	8.6	8.1
0.25	0.013	34.7	9.4	8.9	0.31	0.013	33.6	7.9	7.4
0.25	0.014	32.2	8.8	8.2	0.31	0.014	31.2	7.4	6.8
0.25	0.015	30.0	8.2	7.6	0.31	0.015	29.2	6.9	6.3
0.25	0.016	28.2	7.7	7.0	0.31	0.016	27.3	6.4	5.8
0.26	0.009	49.8	13.2	12.9	0.32	0.009	48.3	11.1	10.8
0.26	0.010	44.9	11.9	11.5	0.32	0.010	43.5	10.0	9.6
0.26	0.011	40.8	10.8	10.4	0.32	0.011	39.5	9.1	8.7
0.26	0.012	37.4	9.9	9.5	0.32	0.012	36.2	8.3	7.9
0.26	0.013	34.5	9.2	8.6	0.32	0.013	33.5	7.7	7.2
0.26	0.014	32.0	8.5	7.9	0.32	0.014	31.1	7.1	6.6
0.26	0.015	29.9	7.9	7.3	0.32	0.015	29.0	6.7	6.1
0.26	0.016	28.0	7.5	6.8	0.32	0.016	27.2	6.3	5.6
0.27	0.009	49.6	12.9	12.5	0.33	0.009	48.0	10.8	10.4
0.27	0.010	44.7	11.6	11.2	0.33	0.010	43.2	9.7	9.3
0.27	0.011	40.6	10.5	10.1	0.33	0.011	39.3	8.8	8.4
0.27	0.012	37.2	9.7	9.2	0.33	0.012	36.0	8.1	7.6
0.27	0.013	34.3	8.9	8.4	0.33	0.013	33.3	7.5	7.0
0.27	0.014	31.9	8.3	7.7	0.33	0.014	30.9	6.9	6.4
0.27	0.015	29.8	7.7	7.1	0.33	0.015	28.8	6.5	5.9
0.27	0.016	27.9	7.2	6.6	0.33	0.016	27.0	6.1	5.4
0.28	0.009	49.4	12.5	12.2	0.34	0.009	47.8	10.5	10.1
0.28	0.010	44.4	11.3	10.9	0.34	0.010	43.0	9.4	9.0
0.28	0.011	40.4	10.2	9.8	0.34	0.011	39.1	8.6	8.1
0.28	0.012	37.0	9.4	8.9	0.34	0.012	35.8	7.8	7.4
0.28	0.013	34.2	8.7	8.1	0.34	0.013	33.1	7.2	6.7
0.28	0.014	31.7	8.0	7.5	0.34	0.014	30.7	6.7	6.2
0.28	0.015	29.6	7.5	6.9	0.34	0.015	28.7	6.3	5.7
0.28	0.016	27.8	7.0	6.4	0.34	0.016	26.9	5.9	5.3
0.29	0.009	49.1	12.2	11.8	0.35	0.009	47.5	10.1	9.8
0.29	0.010	44.2	10.9	10.5	0.35	0.010	42.7	9.1	8.7
0.29	0.011	40.2	9.9	9.5	0.35	0.011	38.8	8.3	7.9
0.29	0.012	36.8	9.1	8.6	0.35	0.012	35.6	7.6	7.1
0.29	0.013	34.0	8.4	7.9	0.35	0.013	32.9	7.0	6.5
0.29	0.014	31.6	7.8	7.3	0.35	0.014	30.5	6.5	6.0
0.29	0.015	29.5	7.3	6.7	0.35	0.015	28.5	6.1	5.5
0.29	0.016	27.6	6.8	6.2	0.35	0.016	26.7	5.7	5.1
0.30	0.009	48.9	11.8	11.5	0.36	0.009	47.1	9.8	9.5
0.30	0.010	44.0	10.6	10.2	0.36	0.010	42.4	8.8	8.5
0.30	0.011	40.0	9.7	9.2	0.36	0.011	38.6	8.0	7.6
0.30	0.012	36.7	8.9	8.4	0.36	0.012	35.4	7.4	6.9
0.30	0.013	33.8	8.2	7.7	0.36	0.013	32.6	6.8	6.3
0.30	0.014	31.4	7.6	7.0	0.36	0.014	30.3	6.3	5.8
0.30	0.015	29.3	7.1	6.5	0.36	0.015	28.3	5.9	5.3
0.30	0.016	27.5	6.6	6.0	0.36	0.016	26.5	5.5	4.9

Chart A-1.10 (Continued): Only For Barrels With An Inside Diameter Of 1.10 Inches

(Arbor diameters of from 0.37 Inches thru 0.48 Inches, and spring thicknesses of from 0.009 Thru 0.016 Inches)

For An Arbor Diameter Of	And A Spring Thickness Of	The BEST Spring Length Is	Max Turns Using BEST Length	Max Turns If Spring Is 10" Short Or Long	For An Arbor Diameter Of	And A Spring Thickness Of	The BEST Spring Length Is	Max Turns Using BEST Length	Max Turns If Spring Is 10" Short Or Long
0.37	0.009	46.8	9.5	9.2	0.43	0.009	44.7	7.8	7.5
0.37	0.010	42.1	8.6	8.2	0.43	0.010	40.3	7.0	6.7
0.37	0.011	38.3	7.8	7.4	0.43	0.011	36.6	6.4	6.0
0.37	0.012	35.1	7.1	6.7	0.43	0.012	33.5	5.8	5.4
0.37	0.013	32.4	6.6	6.1	0.43	0.013	31.0	5.4	4.9
0.37	0.014	30.1	6.1	5.6	0.43	0.014	28.8	5.0	4.5
0.37	0.015	28.1	5.7	5.1	0.43	0.015	26.8	4.7	4.1
0.37	0.016	26.3	5.4	4.7	0.43	0.016	25.2	4.4	3.8
0.38	0.009	46.5	9.2	8.9	0.44	0.009	44.3	7.5	7.2
0.38	0.010	41.8	8.3	7.9	0.44	0.010	39.9	6.8	6.4
0.38	0.011	38.0	7.5	7.1	0.44	0.011	36.3	6.2	5.8
0.38	0.012	34.9	6.9	6.5	0.44	0.012	33.3	5.6	5.2
0.38	0.013	32.2	6.4	5.9	0.44	0.013	30.7	5.2	4.8
0.38	0.014	29.9	5.9	5.4	0.44	0.014	28.5	4.8	4.3
0.38	0.015	27.9	5.5	5.0	0.44	0.015	26.6	4.5	4.0
0.38	0.016	26.2	5.2	4.6	0.44	0.016	24.9	4.2	3.7
0.39	0.009	46.2	8.9	8.6	0.45	0.009	44.0	7.3	7.0
0.39	0.010	41.5	8.0	7.7	0.45	0.010	39.6	6.5	6.2
0.39	0.011	37.8	7.3	6.9	0.45	0.011	36.0	5.9	5.6
0.39	0.012	34.6	6.7	6.2	0.45	0.012	33.0	5.4	5.0
0.39	0.013	32.0	6.2	5.7	0.45	0.013	30.4	5.0	4.6
0.39	0.014	29.7	5.7	5.2	0.45	0.014	28.3	4.7	4.2
0.39	0.015	27.7	5.4	4.8	0.45	0.015	26.4	4.4	3.8
0.39	0.016	26.0	5.0	4.4	0.45	0.016	24.7	4.1	3.5
0.40	0.009	45.8	8.6	8.3	0.46	0.009	43.6	7.0	6.7
0.40	0.010	41.2	7.8	7.4	0.46	0.010	39.2	6.3	6.0
0.40	0.011	37.5	7.1	6.7	0.46	0.011	35.6	5.7	5.4
0.40	0.012	34.4	6.5	6.0	0.46	0.012	32.7	5.3	4.8
0.40	0.013	31.7	6.0	5.5	0.46	0.013	30.2	4.9	4.4
0.40	0.014	29.5	5.5	5.0	0.46	0.014	28.0	4.5	4.0
0.40	0.015	27.5	5.2	4.6	0.46	0.015	26.1	4.2	3.7
0.40	0.016	25.8	4.9	4.3	0.46	0.016	24.5	3.9	3.4
0.41	0.009	45.5	8.3	8.0	0.47	0.009	43.2	6.8	6.5
0.41	0.010	40.9	7.5	7.2	0.47	0.010	38.8	6.1	5.7
0.41	0.011	37.2	6.8	6.4	0.47	0.011	35.3	5.5	5.2
0.41	0.012	34.1	6.3	5.8	0.47	0.012	32.4	5.1	4.7
0.41	0.013	31.5	5.8	5.3	0.47	0.013	29.9	4.7	4.2
0.41	0.014	29.2	5.4	4.9	0.47	0.014	27.7	4.3	3.9
0.41	0.015	27.3	5.0	4.5	0.47	0.015	25.9	4.1	3.5
0.41	0.016	25.6	4.7	4.1	0.47	0.016	24.3	3.8	3.3
0.42	0.009	45.1	8.1	7.7	0.48	0.009	42.7	6.5	6.2
0.42	0.010	40.6	7.3	6.9	0.48	0.010	38.5	5.9	5.5
0.42	0.011	36.9	6.6	6.2	0.48	0.011	35.0	5.3	5.0
0.42	0.012	33.8	6.0	5.6	0.48	0.012	32.1	4.9	4.5
0.42	0.013	31.2	5.6	5.1	0.48	0.013	29.6	4.5	4.1
0.42	0.014	29.0	5.2	4.7	0.48	0.014	27.5	4.2	3.7
0.42	0.015	27.1	4.8	4.3	0.48	0.015	25.6	3.9	3.4
0.42	0.016	25.4	4.5	4.0	0.48	0.016	24.0	3.7	3.1

Selecting the Perfect Spring for Your Barreled Spring Clock

Chart A-1.15: Only For Barrels With An Inside Diameter Of 1.15 Inches

(Arbor diameters of from 0.25 Inches thru 0.36 Inches, and spring thicknesses of from 0.009 Thru 0.016 Inches)

For An Arbor Diameter Of	And A Spring Thickness Of	The BEST Spring Length Is	Max Turns Using BEST Length	Max Turns If Spring Is 10" Short Or Long	For An Arbor Diameter Of	And A Spring Thickness Of	The BEST Spring Length Is	Max Turns Using BEST Length	Max Turns If Spring Is 10" Short Or Long
0.25	0.009	55.0	14.7	14.4	0.31	0.009	53.5	12.5	12.2
0.25	0.010	49.5	13.2	12.9	0.31	0.010	48.2	11.2	10.9
0.25	0.011	45.0	12.0	11.6	0.31	0.011	43.8	10.2	9.8
0.25	0.012	41.2	11.0	10.6	0.31	0.012	40.1	9.3	8.9
0.25	0.013	38.1	10.2	9.7	0.31	0.013	37.0	8.6	8.2
0.25	0.014	35.3	9.4	8.9	0.31	0.014	34.4	8.0	7.5
0.25	0.015	33.0	8.8	8.3	0.31	0.015	32.1	7.5	7.0
0.25	0.016	30.9	8.3	7.7	0.31	0.016	30.1	7.0	6.5
0.26	0.009	54.8	14.3	14.0	0.32	0.009	53.2	12.1	11.8
0.26	0.010	49.3	12.9	12.5	0.32	0.010	47.9	10.9	10.6
0.26	0.011	44.8	11.7	11.3	0.32	0.011	43.6	9.9	9.5
0.26	0.012	41.1	10.7	10.3	0.32	0.012	39.9	9.1	8.7
0.26	0.013	37.9	9.9	9.4	0.32	0.013	36.9	8.4	7.9
0.26	0.014	35.2	9.2	8.7	0.32	0.014	34.2	7.8	7.3
0.26	0.015	32.9	8.6	8.0	0.32	0.015	31.9	7.3	6.8
0.26	0.016	30.8	8.0	7.5	0.32	0.016	29.9	6.8	6.3
0.27	0.009	54.5	13.9	13.6	0.33	0.009	53.0	11.8	11.5
0.27	0.010	49.1	12.5	12.2	0.33	0.010	47.7	10.6	10.3
0.27	0.011	44.6	11.4	11.0	0.33	0.011	43.3	9.6	9.3
0.27	0.012	40.9	10.4	10.0	0.33	0.012	39.7	8.8	8.4
0.27	0.013	37.7	9.6	9.2	0.33	0.013	36.7	8.2	7.7
0.27	0.014	35.1	8.9	8.5	0.33	0.014	34.0	7.6	7.1
0.27	0.015	32.7	8.4	7.8	0.33	0.015	31.8	7.1	6.6
0.27	0.016	30.7	7.8	7.3	0.33	0.016	29.8	6.6	6.1
0.28	0.009	54.3	13.5	13.2	0.34	0.009	52.7	11.4	11.1
0.28	0.010	48.9	12.2	11.8	0.34	0.010	47.4	10.3	10.0
0.28	0.011	44.4	11.1	10.7	0.34	0.011	43.1	9.4	9.0
0.28	0.012	40.7	10.2	9.7	0.34	0.012	39.5	8.6	8.2
0.28	0.013	37.6	9.4	8.9	0.34	0.013	36.5	7.9	7.5
0.28	0.014	34.9	8.7	8.2	0.34	0.014	33.9	7.4	6.9
0.28	0.015	32.6	8.1	7.6	0.34	0.015	31.6	6.9	6.4
0.28	0.016	30.5	7.6	7.1	0.34	0.016	29.6	6.4	5.9
0.29	0.009	54.0	13.2	12.9	0.35	0.009	52.4	11.1	10.8
0.29	0.010	48.6	11.9	11.5	0.35	0.010	47.1	10.0	9.7
0.29	0.011	44.2	10.8	10.4	0.35	0.011	42.8	9.1	8.7
0.29	0.012	40.5	9.9	9.5	0.35	0.012	39.3	8.3	7.9
0.29	0.013	37.4	9.1	8.7	0.35	0.013	36.2	7.7	7.3
0.29	0.014	34.7	8.5	8.0	0.35	0.014	33.7	7.1	6.7
0.29	0.015	32.4	7.9	7.4	0.35	0.015	31.4	6.7	6.2
0.29	0.016	30.4	7.4	6.8	0.35	0.016	29.5	6.3	5.7
0.30	0.009	53.8	12.8	12.5	0.36	0.009	52.1	10.8	10.5
0.30	0.010	48.4	11.5	11.2	0.36	0.010	46.8	9.7	9.4
0.30	0.011	44.0	10.5	10.1	0.36	0.011	42.6	8.8	8.5
0.30	0.012	40.3	9.6	9.2	0.36	0.012	39.0	8.1	7.7
0.30	0.013	37.2	8.9	8.4	0.36	0.013	36.0	7.5	7.0
0.30	0.014	34.6	8.2	7.8	0.36	0.014	33.5	6.9	6.5
0.30	0.015	32.3	7.7	7.2	0.36	0.015	31.2	6.5	6.0
0.30	0.016	30.3	7.2	6.7	0.36	0.016	29.3	6.1	5.5

Chart A-1.15 (Continued): Only For Barrels With An Inside Diameter Of 1.15 Inches

(Arbor diameters of from 0.37 Inches thru 0.48 Inches, and spring thicknesses of from 0.009 Thru 0.016 Inches)

For An Arbor Diameter Of	And A Spring Thickness Of	The BEST Spring Length Is	Max Turns Using BEST Length	Max Turns If Spring Is 10" Short Or Long	For An Arbor Diameter Of	And A Spring Thickness Of	The BEST Spring Length Is	Max Turns Using BEST Length	Max Turns If Spring Is 10" Short Or Long
0.37	0.009	51.7	10.5	10.2	0.43	0.009	49.6	8.7	8.4
0.37	0.010	46.6	9.4	9.1	0.43	0.010	44.7	7.8	7.5
0.37	0.011	42.3	8.6	8.2	0.43	0.011	40.6	7.1	6.8
0.37	0.012	38.8	7.9	7.5	0.43	0.012	37.2	6.5	6.1
0.37	0.013	35.8	7.2	6.8	0.43	0.013	34.4	6.0	5.6
0.37	0.014	33.3	6.7	6.3	0.43	0.014	31.9	5.6	5.1
0.37	0.015	31.0	6.3	5.8	0.43	0.015	29.8	5.2	4.7
0.37	0.016	29.1	5.9	5.4	0.43	0.016	27.9	4.9	4.4
0.38	0.009	51.4	10.2	9.9	0.44	0.009	49.3	8.4	8.1
0.38	0.010	46.3	9.1	8.8	0.44	0.010	44.3	7.6	7.3
0.38	0.011	42.1	8.3	8.0	0.44	0.011	40.3	6.9	6.5
0.38	0.012	38.6	7.6	7.2	0.44	0.012	36.9	6.3	5.9
0.38	0.013	35.6	7.0	6.6	0.44	0.013	34.1	5.8	5.4
0.38	0.014	33.0	6.5	6.1	0.44	0.014	31.7	5.4	5.0
0.38	0.015	30.8	6.1	5.6	0.44	0.015	29.6	5.0	4.6
0.38	0.016	28.9	5.7	5.2	0.44	0.016	27.7	4.7	4.2
0.39	0.009	51.1	9.9	9.6	0.45	0.009	48.9	8.1	7.9
0.39	0.010	46.0	8.9	8.5	0.45	0.010	44.0	7.3	7.0
0.39	0.011	41.8	8.1	7.7	0.45	0.011	40.0	6.7	6.3
0.39	0.012	38.3	7.4	7.0	0.45	0.012	36.7	6.1	5.7
0.39	0.013	35.4	6.8	6.4	0.45	0.013	33.8	5.6	5.2
0.39	0.014	32.8	6.3	5.9	0.45	0.014	31.4	5.2	4.8
0.39	0.015	30.6	5.9	5.4	0.45	0.015	29.3	4.9	4.4
0.39	0.016	28.7	5.5	5.0	0.45	0.016	27.5	4.6	4.1
0.40	0.009	50.7	9.6	9.3	0.46	0.009	48.5	7.9	7.6
0.40	0.010	45.7	8.6	8.3	0.46	0.010	43.6	7.1	6.8
0.40	0.011	41.5	7.8	7.5	0.46	0.011	39.7	6.4	6.1
0.40	0.012	38.0	7.2	6.8	0.46	0.012	36.4	5.9	5.5
0.40	0.013	35.1	6.6	6.2	0.46	0.013	33.6	5.4	5.0
0.40	0.014	32.6	6.1	5.7	0.46	0.014	31.2	5.1	4.6
0.40	0.015	30.4	5.7	5.2	0.46	0.015	29.1	4.7	4.3
0.40	0.016	28.5	5.4	4.9	0.46	0.016	27.3	4.4	3.9
0.41	0.009	50.4	9.3	9.0	0.47	0.009	48.1	7.6	7.3
0.41	0.010	45.3	8.3	8.0	0.47	0.010	43.3	6.8	6.5
0.41	0.011	41.2	7.6	7.2	0.47	0.011	39.3	6.2	5.9
0.41	0.012	37.8	6.9	6.6	0.47	0.012	36.0	5.7	5.3
0.41	0.013	34.9	6.4	6.0	0.47	0.013	33.3	5.3	4.9
0.41	0.014	32.4	6.0	5.5	0.47	0.014	30.9	4.9	4.5
0.41	0.015	30.2	5.6	5.1	0.47	0.015	28.8	4.6	4.1
0.41	0.016	28.3	5.2	4.7	0.47	0.016	27.0	4.3	3.8
0.42	0.009	50.0	9.0	8.7	0.48	0.009	47.7	7.4	7.1
0.42	0.010	45.0	8.1	7.8	0.48	0.010	42.9	6.6	6.3
0.42	0.011	40.9	7.3	7.0	0.48	0.011	39.0	6.0	5.7
0.42	0.012	37.5	6.7	6.3	0.48	0.012	35.7	5.5	5.2
0.42	0.013	34.6	6.2	5.8	0.48	0.013	33.0	5.1	4.7
0.42	0.014	32.1	5.8	5.3	0.48	0.014	30.6	4.7	4.3
0.42	0.015	30.0	5.4	4.9	0.48	0.015	28.6	4.4	4.0
0.42	0.016	28.1	5.0	4.5	0.48	0.016	26.8	4.1	3.7

Chart A-1.20: Only For Barrels With An Inside Diameter Of 1.20 Inches

(Arbor diameters of from 0.26 Inches thru 0.37 Inches, and spring thicknesses of from 0.009 Thru 0.016 Inches)

For An Arbor Diameter Of	And A Spring Thickness Of	The BEST Spring Length Is	Max Turns Using BEST Length	Max Turns If Spring Is 10" Short Or Long	For An Arbor Diameter Of	And A Spring Thickness Of	The BEST Spring Length Is	Max Turns Using BEST Length	Max Turns If Spring Is 10" Short Or Long
0.26	0.009	59.9	15.4	15.1	0.32	0.009	58.4	13.1	12.9
0.26	0.010	53.9	13.8	13.5	0.32	0.010	52.5	11.8	11.5
0.26	0.011	49.0	12.6	12.2	0.32	0.011	47.8	10.7	10.4
0.26	0.012	44.9	11.5	11.1	0.32	0.012	43.8	9.8	9.5
0.26	0.013	41.5	10.6	10.2	0.32	0.013	40.4	9.1	8.7
0.26	0.014	38.5	9.9	9.4	0.32	0.014	37.5	8.4	8.0
0.26	0.015	35.9	9.2	8.7	0.32	0.015	35.0	7.9	7.4
0.26	0.016	33.7	8.6	8.1	0.32	0.016	32.8	7.4	6.9
0.27	0.009	59.7	15.0	14.7	0.33	0.009	58.1	12.8	12.5
0.27	0.010	53.7	13.5	13.2	0.33	0.010	52.3	11.5	11.2
0.27	0.011	48.8	12.2	11.9	0.33	0.011	47.5	10.5	10.1
0.27	0.012	44.7	11.2	10.9	0.33	0.012	43.6	9.6	9.2
0.27	0.013	41.3	10.4	10.0	0.33	0.013	40.2	8.8	8.5
0.27	0.014	38.3	9.6	9.2	0.33	0.014	37.3	8.2	7.8
0.27	0.015	35.8	9.0	8.5	0.33	0.015	34.8	7.7	7.2
0.27	0.016	33.6	8.4	7.9	0.33	0.016	32.7	7.2	6.7
0.28	0.009	59.4	14.6	14.3	0.34	0.009	57.8	12.4	12.2
0.28	0.010	53.5	13.1	12.8	0.34	0.010	52.0	11.2	10.9
0.28	0.011	48.6	11.9	11.6	0.34	0.011	47.3	10.2	9.8
0.28	0.012	44.6	10.9	10.6	0.34	0.012	43.3	9.3	9.0
0.28	0.013	41.1	10.1	9.7	0.34	0.013	40.0	8.6	8.2
0.28	0.014	38.2	9.4	8.9	0.34	0.014	37.1	8.0	7.6
0.28	0.015	35.6	8.8	8.3	0.34	0.015	34.7	7.5	7.0
0.28	0.016	33.4	8.2	7.7	0.34	0.016	32.5	7.0	6.5
0.29	0.009	59.2	14.2	13.9	0.35	0.009	57.5	12.1	11.8
0.29	0.010	53.2	12.8	12.5	0.35	0.010	51.7	10.9	10.6
0.29	0.011	48.4	11.6	11.3	0.35	0.011	47.0	9.9	9.6
0.29	0.012	44.4	10.7	10.3	0.35	0.012	43.1	9.1	8.7
0.29	0.013	41.0	9.8	9.4	0.35	0.013	39.8	8.4	8.0
0.29	0.014	38.0	9.1	8.7	0.35	0.014	37.0	7.8	7.4
0.29	0.015	35.5	8.5	8.1	0.35	0.015	34.5	7.3	6.8
0.29	0.016	33.3	8.0	7.5	0.35	0.016	32.3	6.8	6.3
0.30	0.009	58.9	13.8	13.6	0.36	0.009	57.2	11.8	11.5
0.30	0.010	53.0	12.5	12.2	0.36	0.010	51.5	10.6	10.3
0.30	0.011	48.2	11.3	11.0	0.36	0.011	46.8	9.6	9.3
0.30	0.012	44.2	10.4	10.0	0.36	0.012	42.9	8.8	8.5
0.30	0.013	40.8	9.6	9.2	0.36	0.013	39.6	8.1	7.8
0.30	0.014	37.9	8.9	8.5	0.36	0.014	36.8	7.6	7.1
0.30	0.015	35.3	8.3	7.8	0.36	0.015	34.3	7.1	6.6
0.30	0.016	33.1	7.8	7.3	0.36	0.016	32.2	6.6	6.1
0.31	0.009	58.6	13.5	13.2	0.37	0.009	56.9	11.4	11.2
0.31	0.010	52.8	12.1	11.8	0.37	0.010	51.2	10.3	10.0
0.31	0.011	48.0	11.0	10.7	0.37	0.011	46.5	9.4	9.0
0.31	0.012	44.0	10.1	9.7	0.37	0.012	42.6	8.6	8.2
0.31	0.013	40.6	9.3	8.9	0.37	0.013	39.4	7.9	7.5
0.31	0.014	37.7	8.7	8.2	0.37	0.014	36.6	7.4	6.9
0.31	0.015	35.2	8.1	7.6	0.37	0.015	34.1	6.9	6.4
0.31	0.016	33.0	7.6	7.1	0.37	0.016	32.0	6.4	6.0

Chart A-1.20 (Continued): Only For Barrels With An Inside Diameter Of 1.20 Inches

(Arbor diameters of from 0.38 Inches thru 0.49 Inches, and spring thicknesses of from 0.009 Thru 0.016 Inches)

For An Arbor Diameter Of	And A Spring Thickness Of	The BEST Spring Length Is	Max Turns Using BEST Length	Max Turns If Spring Is 10" Short Or Long	For An Arbor Diameter Of	And A Spring Thickness Of	The BEST Spring Length Is	Max Turns Using BEST Length	Max Turns If Spring Is 10" Short Or Long
0.38	0.009	56.5	11.1	10.9	0.44	0.009	54.4	9.3	9.1
0.38	0.010	50.9	10.0	9.7	0.44	0.010	48.9	8.4	8.1
0.38	0.011	46.3	9.1	8.8	0.44	0.011	44.5	7.6	7.3
0.38	0.012	42.4	8.3	8.0	0.44	0.012	40.8	7.0	6.6
0.38	0.013	39.1	7.7	7.3	0.44	0.013	37.7	6.4	6.1
0.38	0.014	36.3	7.1	6.7	0.44	0.014	35.0	6.0	5.6
0.38	0.015	33.9	6.7	6.2	0.44	0.015	32.6	5.6	5.2
0.38	0.016	31.8	6.3	5.8	0.44	0.016	30.6	5.2	4.8
0.39	0.009	56.2	10.8	10.5	0.45	0.009	54.0	9.0	8.8
0.39	0.010	50.6	9.7	9.4	0.45	0.010	48.6	8.1	7.8
0.39	0.011	46.0	8.8	8.5	0.45	0.011	44.2	7.4	7.1
0.39	0.012	42.1	8.1	7.8	0.45	0.012	40.5	6.8	6.4
0.39	0.013	38.9	7.5	7.1	0.45	0.013	37.4	6.2	5.9
0.39	0.014	36.1	6.9	6.5	0.45	0.014	34.7	5.8	5.4
0.39	0.015	33.7	6.5	6.0	0.45	0.015	32.4	5.4	5.0
0.39	0.016	31.6	6.1	5.6	0.45	0.016	30.4	5.1	4.6
0.40	0.009	55.9	10.5	10.2	0.46	0.009	53.6	8.7	8.5
0.40	0.010	50.3	9.4	9.2	0.46	0.010	48.2	7.9	7.6
0.40	0.011	45.7	8.6	8.3	0.46	0.011	43.9	7.2	6.9
0.40	0.012	41.9	7.9	7.5	0.46	0.012	40.2	6.6	6.2
0.40	0.013	38.7	7.3	6.9	0.46	0.013	37.1	6.1	5.7
0.40	0.014	35.9	6.7	6.3	0.46	0.014	34.5	5.6	5.2
0.40	0.015	33.5	6.3	5.9	0.46	0.015	32.2	5.2	4.8
0.40	0.016	31.4	5.9	5.4	0.46	0.016	30.1	4.9	4.5
0.41	0.009	55.5	10.2	9.9	0.47	0.009	53.2	8.5	8.2
0.41	0.010	49.9	9.2	8.9	0.47	0.010	47.9	7.6	7.4
0.41	0.011	45.4	8.3	8.0	0.47	0.011	43.5	6.9	6.6
0.41	0.012	41.6	7.6	7.3	0.47	0.012	39.9	6.4	6.0
0.41	0.013	38.4	7.1	6.7	0.47	0.013	36.8	5.9	5.5
0.41	0.014	35.7	6.5	6.1	0.47	0.014	34.2	5.4	5.1
0.41	0.015	33.3	6.1	5.7	0.47	0.015	31.9	5.1	4.7
0.41	0.016	31.2	5.7	5.3	0.47	0.016	29.9	4.8	4.3
0.42	0.009	55.1	9.9	9.6	0.48	0.009	52.8	8.2	8.0
0.42	0.010	49.6	8.9	8.6	0.48	0.010	47.5	7.4	7.1
0.42	0.011	45.1	8.1	7.8	0.48	0.011	43.2	6.7	6.4
0.42	0.012	41.4	7.4	7.1	0.48	0.012	39.6	6.2	5.8
0.42	0.013	38.2	6.8	6.5	0.48	0.013	36.5	5.7	5.3
0.42	0.014	35.4	6.4	6.0	0.48	0.014	33.9	5.3	4.9
0.42	0.015	33.1	5.9	5.5	0.48	0.015	31.7	4.9	4.5
0.42	0.016	31.0	5.6	5.1	0.48	0.016	29.7	4.6	4.2
0.43	0.009	54.8	9.6	9.3	0.49	0.009	52.4	7.9	7.7
0.43	0.010	49.3	8.6	8.4	0.49	0.010	47.1	7.2	6.9
0.43	0.011	44.8	7.9	7.5	0.49	0.011	42.8	6.5	6.2
0.43	0.012	41.1	7.2	6.9	0.49	0.012	39.3	6.0	5.6
0.43	0.013	37.9	6.6	6.3	0.49	0.013	36.2	5.5	5.2
0.43	0.014	35.2	6.2	5.8	0.49	0.014	33.7	5.1	4.7
0.43	0.015	32.9	5.8	5.3	0.49	0.015	31.4	4.8	4.4
0.43	0.016	30.8	5.4	4.9	0.49	0.016	29.4	4.5	4.0

Chart A-1.25: Only For Barrels With An Inside Diameter Of 1.25 Inches

(Arbor diameters of from 0.26 Inches thru 0.37 Inches, and spring thicknesses of from 0.009 Thru 0.016 Inches)

For An Arbor Diameter Of	And A Spring Thickness Of	The BEST Spring Length Is	Max Turns Using BEST Length	Max Turns If Spring Is 10" Short Or Long	For An Arbor Diameter Of	And A Spring Thickness Of	The BEST Spring Length Is	Max Turns Using BEST Length	Max Turns If Spring Is 10" Short Or Long
0.26	0.009	65.2	16.4	16.2	0.32	0.009	63.7	14.2	13.9
0.26	0.010	58.7	14.8	14.5	0.32	0.010	57.3	12.7	12.5
0.26	0.011	53.4	13.4	13.1	0.32	0.011	52.1	11.6	11.3
0.26	0.012	48.9	12.3	12.0	0.32	0.012	47.8	10.6	10.3
0.26	0.013	45.2	11.4	11.0	0.32	0.013	44.1	9.8	9.4
0.26	0.014	41.9	10.6	10.2	0.32	0.014	41.0	9.1	8.7
0.26	0.015	39.1	9.9	9.4	0.32	0.015	38.2	8.5	8.1
0.26	0.016	36.7	9.2	8.8	0.32	0.016	35.8	8.0	7.5
0.27	0.009	65.0	16.0	15.8	0.33	0.009	63.4	13.8	13.6
0.27	0.010	58.5	14.4	14.2	0.33	0.010	57.1	12.4	12.1
0.27	0.011	53.2	13.1	12.8	0.33	0.011	51.9	11.3	11.0
0.27	0.012	48.7	12.0	11.7	0.33	0.012	47.6	10.3	10.0
0.27	0.013	45.0	11.1	10.7	0.33	0.013	43.9	9.6	9.2
0.27	0.014	41.8	10.3	9.9	0.33	0.014	40.8	8.9	8.5
0.27	0.015	39.0	9.6	9.2	0.33	0.015	38.1	8.3	7.9
0.27	0.016	36.6	9.0	8.6	0.33	0.016	35.7	7.8	7.3
0.28	0.009	64.8	15.6	15.4	0.34	0.009	63.1	13.4	13.2
0.28	0.010	58.3	14.1	13.8	0.34	0.010	56.8	12.1	11.8
0.28	0.011	53.0	12.8	12.5	0.34	0.011	51.7	11.0	10.7
0.28	0.012	48.6	11.7	11.4	0.34	0.012	47.3	10.1	9.8
0.28	0.013	44.8	10.8	10.5	0.34	0.013	43.7	9.3	9.0
0.28	0.014	41.6	10.1	9.7	0.34	0.014	40.6	8.6	8.3
0.28	0.015	38.9	9.4	9.0	0.34	0.015	37.9	8.1	7.7
0.28	0.016	36.4	8.8	8.4	0.34	0.016	35.5	7.6	7.1
0.29	0.009	64.5	15.3	15.0	0.35	0.009	62.8	13.1	12.9
0.29	0.010	58.1	13.7	13.5	0.35	0.010	56.5	11.8	11.5
0.29	0.011	52.8	12.5	12.2	0.35	0.011	51.4	10.7	10.4
0.29	0.012	48.4	11.4	11.1	0.35	0.012	47.1	9.8	9.5
0.29	0.013	44.7	10.6	10.2	0.35	0.013	43.5	9.1	8.7
0.29	0.014	41.5	9.8	9.4	0.35	0.014	40.4	8.4	8.0
0.29	0.015	38.7	9.2	8.7	0.35	0.015	37.7	7.9	7.5
0.29	0.016	36.3	8.6	8.1	0.35	0.016	35.3	7.4	6.9
0.30	0.009	64.2	14.9	14.6	0.36	0.009	62.5	12.8	12.5
0.30	0.010	57.8	13.4	13.1	0.36	0.010	56.3	11.5	11.2
0.30	0.011	52.6	12.2	11.9	0.36	0.011	51.2	10.4	10.1
0.30	0.012	48.2	11.2	10.8	0.36	0.012	46.9	9.6	9.3
0.30	0.013	44.5	10.3	10.0	0.36	0.013	43.3	8.8	8.5
0.30	0.014	41.3	9.6	9.2	0.36	0.014	40.2	8.2	7.8
0.30	0.015	38.5	8.9	8.5	0.36	0.015	37.5	7.7	7.3
0.30	0.016	36.1	8.4	7.9	0.36	0.016	35.2	7.2	6.8
0.31	0.009	64.0	14.5	14.3	0.37	0.009	62.2	12.4	12.2
0.31	0.010	57.6	13.1	12.8	0.37	0.010	56.0	11.2	10.9
0.31	0.011	52.4	11.9	11.6	0.37	0.011	50.9	10.2	9.9
0.31	0.012	48.0	10.9	10.6	0.37	0.012	46.7	9.3	9.0
0.31	0.013	44.3	10.1	9.7	0.37	0.013	43.1	8.6	8.3
0.31	0.014	41.1	9.3	9.0	0.37	0.014	40.0	8.0	7.6
0.31	0.015	38.4	8.7	8.3	0.37	0.015	37.3	7.5	7.1
0.31	0.016	36.0	8.2	7.7	0.37	0.016	35.0	7.0	6.6

Chart A-1.25 (Continued): Only For Barrels With An Inside Diameter Of 1.25 Inches

(Arbor diameters of from 0.38 Inches thru 0.49 Inches, and spring thicknesses of from 0.009 Thru 0.016 Inches)

For An Arbor Diameter Of	And A Spring Thickness Of	The BEST Spring Length Is	Max Turns Using BEST Length	Max Turns If Spring Is 10" Short Or Long	For An Arbor Diameter Of	And A Spring Thickness Of	The BEST Spring Length Is	Max Turns Using BEST Length	Max Turns If Spring Is 10" Short Or Long
0.38	0.009	61.9	12.1	11.9	0.44	0.009	59.7	10.2	10.0
0.38	0.010	55.7	10.9	10.6	0.44	0.010	53.8	9.2	9.0
0.38	0.011	50.6	9.9	9.6	0.44	0.011	48.9	8.4	8.1
0.38	0.012	46.4	9.1	8.8	0.44	0.012	44.8	7.7	7.4
0.38	0.013	42.8	8.4	8.0	0.44	0.013	41.4	7.1	6.8
0.38	0.014	39.8	7.8	7.4	0.44	0.014	38.4	6.6	6.2
0.38	0.015	37.1	7.3	6.9	0.44	0.015	35.8	6.1	5.8
0.38	0.016	34.8	6.8	6.4	0.44	0.016	33.6	5.8	5.4
0.39	0.009	61.5	11.8	11.5	0.45	0.009	59.3	9.9	9.7
0.39	0.010	55.4	10.6	10.3	0.45	0.010	53.4	8.9	8.7
0.39	0.011	50.4	9.6	9.3	0.45	0.011	48.6	8.1	7.9
0.39	0.012	46.2	8.8	8.5	0.45	0.012	44.5	7.5	7.2
0.39	0.013	42.6	8.1	7.8	0.45	0.013	41.1	6.9	6.6
0.39	0.014	39.6	7.6	7.2	0.45	0.014	38.1	6.4	6.0
0.39	0.015	36.9	7.1	6.7	0.45	0.015	35.6	6.0	5.6
0.39	0.016	34.6	6.6	6.2	0.45	0.016	33.4	5.6	5.2
0.40	0.009	61.2	11.4	11.2	0.46	0.009	58.9	9.6	9.4
0.40	0.010	55.1	10.3	10.0	0.46	0.010	53.0	8.7	8.4
0.40	0.011	50.1	9.4	9.1	0.46	0.011	48.2	7.9	7.6
0.40	0.012	45.9	8.6	8.3	0.46	0.012	44.2	7.2	6.9
0.40	0.013	42.4	7.9	7.6	0.46	0.013	40.8	6.7	6.4
0.40	0.014	39.3	7.4	7.0	0.46	0.014	37.9	6.2	5.9
0.40	0.015	36.7	6.9	6.5	0.46	0.015	35.4	5.8	5.4
0.40	0.016	34.4	6.4	6.0	0.46	0.016	33.2	5.4	5.0
0.41	0.009	60.8	11.1	10.9	0.47	0.009	58.5	9.4	9.1
0.41	0.010	54.8	10.0	9.8	0.47	0.010	52.7	8.4	8.2
0.41	0.011	49.8	9.1	8.8	0.47	0.011	47.9	7.7	7.4
0.41	0.012	45.6	8.4	8.0	0.47	0.012	43.9	7.0	6.7
0.41	0.013	42.1	7.7	7.4	0.47	0.013	40.5	6.5	6.2
0.41	0.014	39.1	7.2	6.8	0.47	0.014	37.6	6.0	5.7
0.41	0.015	36.5	6.7	6.3	0.47	0.015	35.1	5.6	5.3
0.41	0.016	34.2	6.3	5.9	0.47	0.016	32.9	5.3	4.9
0.42	0.009	60.5	10.8	10.6	0.48	0.009	58.1	9.1	8.9
0.42	0.010	54.4	9.7	9.5	0.48	0.010	52.3	8.2	7.9
0.42	0.011	49.5	8.9	8.6	0.48	0.011	47.6	7.4	7.2
0.42	0.012	45.4	8.1	7.8	0.48	0.012	43.6	6.8	6.5
0.42	0.013	41.9	7.5	7.2	0.48	0.013	40.2	6.3	6.0
0.42	0.014	38.9	7.0	6.6	0.48	0.014	37.4	5.8	5.5
0.42	0.015	36.3	6.5	6.1	0.48	0.015	34.9	5.5	5.1
0.42	0.016	34.0	6.1	5.7	0.48	0.016	32.7	5.1	4.7
0.43	0.009	60.1	10.5	10.3	0.49	0.009	57.7	8.8	8.6
0.43	0.010	54.1	9.5	9.2	0.49	0.010	51.9	7.9	7.7
0.43	0.011	49.2	8.6	8.3	0.49	0.011	47.2	7.2	7.0
0.43	0.012	45.1	7.9	7.6	0.49	0.012	43.3	6.6	6.3
0.43	0.013	41.6	7.3	7.0	0.49	0.013	39.9	6.1	5.8
0.43	0.014	38.6	6.8	6.4	0.49	0.014	37.1	5.7	5.3
0.43	0.015	36.1	6.3	5.9	0.49	0.015	34.6	5.3	4.9
0.43	0.016	33.8	5.9	5.5	0.49	0.016	32.5	5.0	4.6

Selecting the Perfect Spring for Your Barreled Spring Clock

Chart A-1.30: Only For Barrels With An Inside Diameter Of 1.30 Inches

(Arbor diameters of from 0.26 Inches thru 0.37 Inches, and spring thicknesses of from 0.009 Thru 0.016 Inches)

For An Arbor Diameter Of	And A Spring Thickness Of	The BEST Spring Length Is	Max Turns Using BEST Length	Max Turns If Spring Is 10" Short Or Long	For An Arbor Diameter Of	And A Spring Thickness Of	The BEST Spring Length Is	Max Turns Using BEST Length	Max Turns If Spring Is 10" Short Or Long
0.26	0.009	70.8	17.5	17.3	0.32	0.009	69.3	15.2	15.0
0.26	0.010	63.7	15.7	15.5	0.32	0.010	62.3	13.7	13.4
0.26	0.011	57.9	14.3	14.0	0.32	0.011	56.7	12.4	12.2
0.26	0.012	53.1	13.1	12.8	0.32	0.012	52.0	11.4	11.1
0.26	0.013	49.0	12.1	11.8	0.32	0.013	48.0	10.5	10.2
0.26	0.014	45.5	11.2	10.9	0.32	0.014	44.5	9.8	9.4
0.26	0.015	42.5	10.5	10.1	0.32	0.015	41.6	9.1	8.7
0.26	0.016	39.8	9.8	9.4	0.32	0.016	39.0	8.5	8.2
0.27	0.009	70.6	17.1	16.9	0.33	0.009	69.0	14.8	14.6
0.27	0.010	63.5	15.4	15.1	0.33	0.010	62.1	13.3	13.1
0.27	0.011	57.7	14.0	13.7	0.33	0.011	56.4	12.1	11.9
0.27	0.012	52.9	12.8	12.5	0.33	0.012	51.7	11.1	10.8
0.27	0.013	48.8	11.8	11.5	0.33	0.013	47.8	10.3	9.9
0.27	0.014	45.4	11.0	10.6	0.33	0.014	44.3	9.5	9.2
0.27	0.015	42.3	10.3	9.9	0.33	0.015	41.4	8.9	8.5
0.27	0.016	39.7	9.6	9.2	0.33	0.016	38.8	8.3	8.0
0.28	0.009	70.3	16.7	16.5	0.34	0.009	68.7	14.5	14.2
0.28	0.010	63.3	15.0	14.8	0.34	0.010	61.8	13.0	12.8
0.28	0.011	57.5	13.7	13.4	0.34	0.011	56.2	11.8	11.6
0.28	0.012	52.7	12.5	12.2	0.34	0.012	51.5	10.8	10.6
0.28	0.013	48.7	11.6	11.2	0.34	0.013	47.6	10.0	9.7
0.28	0.014	45.2	10.7	10.4	0.34	0.014	44.2	9.3	9.0
0.28	0.015	42.2	10.0	9.7	0.34	0.015	41.2	8.7	8.3
0.28	0.016	39.6	9.4	9.0	0.34	0.016	38.6	8.1	7.8
0.29	0.009	70.1	16.3	16.1	0.35	0.009	68.4	14.1	13.9
0.29	0.010	63.1	14.7	14.4	0.35	0.010	61.6	12.7	12.5
0.29	0.011	57.3	13.3	13.1	0.35	0.011	56.0	11.5	11.3
0.29	0.012	52.6	12.2	11.9	0.35	0.012	51.3	10.6	10.3
0.29	0.013	48.5	11.3	11.0	0.35	0.013	47.4	9.8	9.5
0.29	0.014	45.0	10.5	10.1	0.35	0.014	44.0	9.1	8.7
0.29	0.015	42.0	9.8	9.4	0.35	0.015	41.0	8.5	8.1
0.29	0.016	39.4	9.2	8.8	0.35	0.016	38.5	7.9	7.6
0.30	0.009	69.8	15.9	15.7	0.36	0.009	68.1	13.8	13.5
0.30	0.010	62.8	14.3	14.1	0.36	0.010	61.3	12.4	12.1
0.30	0.011	57.1	13.0	12.8	0.36	0.011	55.7	11.3	11.0
0.30	0.012	52.4	11.9	11.7	0.36	0.012	51.1	10.3	10.0
0.30	0.013	48.3	11.0	10.7	0.36	0.013	47.1	9.5	9.2
0.30	0.014	44.9	10.2	9.9	0.36	0.014	43.8	8.8	8.5
0.30	0.015	41.9	9.6	9.2	0.36	0.015	40.9	8.3	7.9
0.30	0.016	39.3	9.0	8.6	0.36	0.016	38.3	7.7	7.4
0.31	0.009	69.5	15.6	15.3	0.37	0.009	67.8	13.4	13.2
0.31	0.010	62.6	14.0	13.8	0.37	0.010	61.0	12.1	11.8
0.31	0.011	56.9	12.7	12.5	0.37	0.011	55.4	11.0	10.7
0.31	0.012	52.2	11.7	11.4	0.37	0.012	50.8	10.1	9.8
0.31	0.013	48.1	10.8	10.5	0.37	0.013	46.9	9.3	9.0
0.31	0.014	44.7	10.0	9.7	0.37	0.014	43.6	8.6	8.3
0.31	0.015	41.7	9.3	9.0	0.37	0.015	40.7	8.0	7.7
0.31	0.016	39.1	8.8	8.4	0.37	0.016	38.1	7.5	7.2

Chart A-1.30 (Continued): Only For Barrels With An Inside Diameter Of 1.30 Inches

(Arbor diameters of from 0.38 Inches thru 0.49 Inches, and spring thicknesses of from 0.009 Thru 0.016 Inches)

For An Arbor Diameter Of	And A Spring Thickness Of	The BEST Spring Length Is	Max Turns Using BEST Length	Max Turns If Spring Is 10" Short Or Long	For An Arbor Diameter Of	And A Spring Thickness Of	The BEST Spring Length Is	Max Turns Using BEST Length	Max Turns If Spring Is 10" Short Or Long
0.38	0.009	67.4	13.1	12.9	0.44	0.009	65.3	11.2	11.0
0.38	0.010	60.7	11.8	11.5	0.44	0.010	58.8	10.0	9.8
0.38	0.011	55.2	10.7	10.4	0.44	0.011	53.4	9.1	8.9
0.38	0.012	50.6	9.8	9.5	0.44	0.012	49.0	8.4	8.1
0.38	0.013	46.7	9.1	8.8	0.44	0.013	45.2	7.7	7.4
0.38	0.014	43.4	8.4	8.1	0.44	0.014	42.0	7.2	6.9
0.38	0.015	40.5	7.8	7.5	0.44	0.015	39.2	6.7	6.4
0.38	0.016	37.9	7.4	7.0	0.44	0.016	36.7	6.3	5.9
0.39	0.009	67.1	12.7	12.5	0.45	0.009	64.9	10.9	10.7
0.39	0.010	60.4	11.5	11.2	0.45	0.010	58.4	9.8	9.6
0.39	0.011	54.9	10.4	10.2	0.45	0.011	53.1	8.9	8.6
0.39	0.012	50.3	9.6	9.3	0.45	0.012	48.7	8.1	7.9
0.39	0.013	46.5	8.8	8.5	0.45	0.013	44.9	7.5	7.2
0.39	0.014	43.1	8.2	7.9	0.45	0.014	41.7	7.0	6.7
0.39	0.015	40.3	7.6	7.3	0.45	0.015	38.9	6.5	6.2
0.39	0.016	37.7	7.2	6.8	0.45	0.016	36.5	6.1	5.8
0.40	0.009	66.8	12.4	12.2	0.46	0.009	64.5	10.6	10.4
0.40	0.010	60.1	11.2	10.9	0.46	0.010	58.1	9.5	9.3
0.40	0.011	54.6	10.2	9.9	0.46	0.011	52.8	8.6	8.4
0.40	0.012	50.1	9.3	9.0	0.46	0.012	48.4	7.9	7.7
0.40	0.013	46.2	8.6	8.3	0.46	0.013	44.7	7.3	7.0
0.40	0.014	42.9	8.0	7.7	0.46	0.014	41.5	6.8	6.5
0.40	0.015	40.1	7.5	7.1	0.46	0.015	38.7	6.3	6.0
0.40	0.016	37.6	7.0	6.6	0.46	0.016	36.3	5.9	5.6
0.41	0.009	66.4	12.1	11.9	0.47	0.009	64.1	10.3	10.1
0.41	0.010	59.8	10.9	10.7	0.47	0.010	57.7	9.2	9.0
0.41	0.011	54.3	9.9	9.6	0.47	0.011	52.4	8.4	8.2
0.41	0.012	49.8	9.1	8.8	0.47	0.012	48.1	7.7	7.4
0.41	0.013	46.0	8.4	8.1	0.47	0.013	44.4	7.1	6.8
0.41	0.014	42.7	7.8	7.5	0.47	0.014	41.2	6.6	6.3
0.41	0.015	39.8	7.3	6.9	0.47	0.015	38.5	6.2	5.8
0.41	0.016	37.4	6.8	6.4	0.47	0.016	36.1	5.8	5.4
0.42	0.009	66.0	11.8	11.6	0.48	0.009	63.7	10.0	9.8
0.42	0.010	59.4	10.6	10.4	0.48	0.010	57.3	9.0	8.8
0.42	0.011	54.0	9.6	9.4	0.48	0.011	52.1	8.2	7.9
0.42	0.012	49.5	8.8	8.6	0.48	0.012	47.8	7.5	7.2
0.42	0.013	45.7	8.2	7.9	0.48	0.013	44.1	6.9	6.6
0.42	0.014	42.5	7.6	7.3	0.48	0.014	40.9	6.4	6.1
0.42	0.015	39.6	7.1	6.7	0.48	0.015	38.2	6.0	5.7
0.42	0.016	37.1	6.6	6.3	0.48	0.016	35.8	5.6	5.3
0.43	0.009	65.7	11.5	11.3	0.49	0.009	63.3	9.7	9.5
0.43	0.010	59.1	10.3	10.1	0.49	0.010	56.9	8.7	8.5
0.43	0.011	53.7	9.4	9.1	0.49	0.011	51.8	7.9	7.7
0.43	0.012	49.3	8.6	8.3	0.49	0.012	47.4	7.3	7.0
0.43	0.013	45.5	7.9	7.6	0.49	0.013	43.8	6.7	6.4
0.43	0.014	42.2	7.4	7.1	0.49	0.014	40.7	6.2	5.9
0.43	0.015	39.4	6.9	6.5	0.49	0.015	38.0	5.8	5.5
0.43	0.016	36.9	6.5	6.1	0.49	0.016	35.6	5.5	5.1

Chart A-1.35: Only For Barrels With An Inside Diameter Of 1.35 Inches

(Arbor diameters of from 0.26 Inches thru 0.37 Inches, and spring thicknesses of from 0.009 Thru 0.016 Inches)

For An Arbor Diameter Of	And A Spring Thickness Of	The BEST Spring Length Is	Max Turns Using BEST Length	Max Turns If Spring Is 10" Short Or Long	For An Arbor Diameter Of	And A Spring Thickness Of	The BEST Spring Length Is	Max Turns Using BEST Length	Max Turns If Spring Is 10" Short Or Long
0.26	0.009	76.6	18.6	18.4	0.32	0.009	75.1	16.2	16.0
0.26	0.010	68.9	16.7	16.5	0.32	0.010	67.5	14.6	14.4
0.26	0.011	62.6	15.2	14.9	0.32	0.011	61.4	13.3	13.0
0.26	0.012	57.4	13.9	13.7	0.32	0.012	56.3	12.2	11.9
0.26	0.013	53.0	12.9	12.6	0.32	0.013	52.0	11.2	11.0
0.26	0.014	49.2	11.9	11.6	0.32	0.014	48.2	10.4	10.1
0.26	0.015	45.9	11.1	10.8	0.32	0.015	45.0	9.7	9.4
0.26	0.016	43.1	10.4	10.1	0.32	0.016	42.2	9.1	8.8
0.27	0.009	76.3	18.2	18.0	0.33	0.009	74.8	15.9	15.7
0.27	0.010	68.7	16.3	16.1	0.33	0.010	67.3	14.3	14.1
0.27	0.011	62.5	14.9	14.6	0.33	0.011	61.2	13.0	12.7
0.27	0.012	57.3	13.6	13.4	0.33	0.012	56.1	11.9	11.6
0.27	0.013	52.9	12.6	12.3	0.33	0.013	51.8	11.0	10.7
0.27	0.014	49.1	11.7	11.4	0.33	0.014	48.1	10.2	9.9
0.27	0.015	45.8	10.9	10.6	0.33	0.015	44.9	9.5	9.2
0.27	0.016	42.9	10.2	9.9	0.33	0.016	42.1	8.9	8.6
0.28	0.009	76.1	17.8	17.6	0.34	0.009	74.5	15.5	15.3
0.28	0.010	68.5	16.0	15.8	0.34	0.010	67.0	13.9	13.7
0.28	0.011	62.3	14.5	14.3	0.34	0.011	60.9	12.7	12.4
0.28	0.012	57.1	13.3	13.1	0.34	0.012	55.9	11.6	11.4
0.28	0.013	52.7	12.3	12.0	0.34	0.013	51.6	10.7	10.4
0.28	0.014	48.9	11.4	11.1	0.34	0.014	47.9	10.0	9.7
0.28	0.015	45.7	10.7	10.3	0.34	0.015	44.7	9.3	9.0
0.28	0.016	42.8	10.0	9.6	0.34	0.016	41.9	8.7	8.4
0.29	0.009	75.9	17.4	17.2	0.35	0.009	74.2	15.1	14.9
0.29	0.010	68.3	15.6	15.4	0.35	0.010	66.8	13.6	13.4
0.29	0.011	62.1	14.2	14.0	0.35	0.011	60.7	12.4	12.1
0.29	0.012	56.9	13.0	12.8	0.35	0.012	55.6	11.3	11.1
0.29	0.013	52.5	12.0	11.7	0.35	0.013	51.4	10.5	10.2
0.29	0.014	48.8	11.2	10.9	0.35	0.014	47.7	9.7	9.4
0.29	0.015	45.5	10.4	10.1	0.35	0.015	44.5	9.1	8.8
0.29	0.016	42.7	9.8	9.4	0.35	0.016	41.7	8.5	8.2
0.30	0.009	75.6	17.0	16.8	0.36	0.009	73.9	14.8	14.6
0.30	0.010	68.0	15.3	15.1	0.36	0.010	66.5	13.3	13.1
0.30	0.011	61.9	13.9	13.7	0.36	0.011	60.4	12.1	11.9
0.30	0.012	56.7	12.7	12.5	0.36	0.012	55.4	11.1	10.8
0.30	0.013	52.3	11.8	11.5	0.36	0.013	51.1	10.2	10.0
0.30	0.014	48.6	10.9	10.6	0.36	0.014	47.5	9.5	9.2
0.30	0.015	45.4	10.2	9.9	0.36	0.015	44.3	8.9	8.5
0.30	0.016	42.5	9.6	9.2	0.36	0.016	41.5	8.3	8.0
0.31	0.009	75.3	16.6	16.4	0.37	0.009	73.5	14.4	14.2
0.31	0.010	67.8	14.9	14.7	0.37	0.010	66.2	13.0	12.8
0.31	0.011	61.6	13.6	13.3	0.37	0.011	60.2	11.8	11.6
0.31	0.012	56.5	12.5	12.2	0.37	0.012	55.2	10.8	10.6
0.31	0.013	52.2	11.5	11.2	0.37	0.013	50.9	10.0	9.7
0.31	0.014	48.4	10.7	10.4	0.37	0.014	47.3	9.3	9.0
0.31	0.015	45.2	10.0	9.6	0.37	0.015	44.1	8.7	8.3
0.31	0.016	42.4	9.3	9.0	0.37	0.016	41.4	8.1	7.8

Chart A-1.35 (Continued): Only For Barrels With An Inside Diameter Of 1.35 Inches

(Arbor diameters of from 0.38 Inches thru 0.49 Inches, and spring thicknesses of from 0.009 Thru 0.016 Inches)

For An Arbor Diameter Of	And A Spring Thickness Of	The BEST Spring Length Is	Max Turns Using BEST Length	Max Turns If Spring Is 10" Short Or Long	For An Arbor Diameter Of	And A Spring Thickness Of	The BEST Spring Length Is	Max Turns Using BEST Length	Max Turns If Spring Is 10" Short Or Long
0.38	0.009	73.2	14.1	13.9	0.44	0.009	71.1	12.1	11.9
0.38	0.010	65.9	12.7	12.5	0.44	0.010	64.0	10.9	10.7
0.38	0.011	59.9	11.5	11.3	0.44	0.011	58.2	9.9	9.7
0.38	0.012	54.9	10.6	10.3	0.44	0.012	53.3	9.1	8.8
0.38	0.013	50.7	9.7	9.5	0.44	0.013	49.2	8.4	8.1
0.38	0.014	47.1	9.0	8.8	0.44	0.014	45.7	7.8	7.5
0.38	0.015	43.9	8.4	8.1	0.44	0.015	42.6	7.3	7.0
0.38	0.016	41.2	7.9	7.6	0.44	0.016	40.0	6.8	6.5
0.39	0.009	72.9	13.7	13.5	0.45	0.009	70.7	11.8	11.6
0.39	0.010	65.6	12.4	12.2	0.45	0.010	63.6	10.6	10.4
0.39	0.011	59.6	11.2	11.0	0.45	0.011	57.8	9.7	9.4
0.39	0.012	54.7	10.3	10.1	0.45	0.012	53.0	8.9	8.6
0.39	0.013	50.5	9.5	9.2	0.45	0.013	48.9	8.2	7.9
0.39	0.014	46.9	8.8	8.5	0.45	0.014	45.4	7.6	7.3
0.39	0.015	43.7	8.2	7.9	0.45	0.015	42.4	7.1	6.8
0.39	0.016	41.0	7.7	7.4	0.45	0.016	39.8	6.6	6.3
0.40	0.009	72.5	13.4	13.2	0.46	0.009	70.3	11.5	11.3
0.40	0.010	65.3	12.1	11.9	0.46	0.010	63.3	10.3	10.2
0.40	0.011	59.4	11.0	10.7	0.46	0.011	57.5	9.4	9.2
0.40	0.012	54.4	10.1	9.8	0.46	0.012	52.7	8.6	8.4
0.40	0.013	50.2	9.3	9.0	0.46	0.013	48.7	8.0	7.7
0.40	0.014	46.6	8.6	8.3	0.46	0.014	45.2	7.4	7.1
0.40	0.015	43.5	8.0	7.7	0.46	0.015	42.2	6.9	6.6
0.40	0.016	40.8	7.5	7.2	0.46	0.016	39.5	6.5	6.1
0.41	0.009	72.2	13.1	12.9	0.47	0.009	69.9	11.2	11.0
0.41	0.010	65.0	11.8	11.6	0.47	0.010	62.9	10.1	9.9
0.41	0.011	59.1	10.7	10.5	0.47	0.011	57.2	9.2	8.9
0.41	0.012	54.1	9.8	9.6	0.47	0.012	52.4	8.4	8.2
0.41	0.013	50.0	9.0	8.8	0.47	0.013	48.4	7.8	7.5
0.41	0.014	46.4	8.4	8.1	0.47	0.014	44.9	7.2	6.9
0.41	0.015	43.3	7.8	7.5	0.47	0.015	41.9	6.7	6.4
0.41	0.016	40.6	7.4	7.0	0.47	0.016	39.3	6.3	6.0
0.42	0.009	71.8	12.7	12.6	0.48	0.009	69.5	10.9	10.7
0.42	0.010	64.6	11.5	11.3	0.48	0.010	62.5	9.8	9.6
0.42	0.011	58.8	10.4	10.2	0.48	0.011	56.8	8.9	8.7
0.42	0.012	53.9	9.6	9.3	0.48	0.012	52.1	8.2	7.9
0.42	0.013	49.7	8.8	8.6	0.48	0.013	48.1	7.5	7.3
0.42	0.014	46.2	8.2	7.9	0.48	0.014	44.7	7.0	6.7
0.42	0.015	43.1	7.6	7.3	0.48	0.015	41.7	6.5	6.2
0.42	0.016	40.4	7.2	6.8	0.48	0.016	39.1	6.1	5.8
0.43	0.009	71.5	12.4	12.2	0.49	0.009	69.0	10.6	10.4
0.43	0.010	64.3	11.2	11.0	0.49	0.010	62.1	9.6	9.4
0.43	0.011	58.5	10.2	9.9	0.49	0.011	56.5	8.7	8.5
0.43	0.012	53.6	9.3	9.1	0.49	0.012	51.8	8.0	7.7
0.43	0.013	49.5	8.6	8.3	0.49	0.013	47.8	7.3	7.1
0.43	0.014	45.9	8.0	7.7	0.49	0.014	44.4	6.8	6.6
0.43	0.015	42.9	7.5	7.2	0.49	0.015	41.4	6.4	6.1
0.43	0.016	40.2	7.0	6.7	0.49	0.016	38.8	6.0	5.7

Chart A-1.40: Only For Barrels With An Inside Diameter Of 1.40 Inches

(Arbor diameters of from 0.26 Inches thru 0.37 Inches, and spring thicknesses of from 0.009 Thru 0.016 Inches)

For An Arbor Diameter Of	And A Spring Thickness Of	The BEST Spring Length Is	Max Turns Using BEST Length	Max Turns If Spring Is 10" Short Or Long	For An Arbor Diameter Of	And A Spring Thickness Of	The BEST Spring Length Is	Max Turns Using BEST Length	Max Turns If Spring Is 10" Short Or Long
0.26	0.009	82.6	19.7	19.5	0.32	0.009	81.1	17.3	17.1
0.26	0.010	74.3	17.7	17.5	0.32	0.010	72.9	15.5	15.4
0.26	0.011	67.6	16.1	15.9	0.32	0.011	66.3	14.1	13.9
0.26	0.012	61.9	14.7	14.5	0.32	0.012	60.8	13.0	12.7
0.26	0.013	57.2	13.6	13.3	0.32	0.013	56.1	12.0	11.7
0.26	0.014	53.1	12.6	12.4	0.32	0.014	52.1	11.1	10.8
0.26	0.015	49.5	11.8	11.5	0.32	0.015	48.6	10.4	10.1
0.26	0.016	46.4	11.1	10.7	0.32	0.016	45.6	9.7	9.4
0.27	0.009	82.3	19.2	19.1	0.33	0.009	80.8	16.9	16.7
0.27	0.010	74.1	17.3	17.1	0.33	0.010	72.7	15.2	15.0
0.27	0.011	67.4	15.7	15.5	0.33	0.011	66.1	13.8	13.6
0.27	0.012	61.8	14.4	14.2	0.33	0.012	60.6	12.7	12.4
0.27	0.013	57.0	13.3	13.1	0.33	0.013	55.9	11.7	11.4
0.27	0.014	52.9	12.4	12.1	0.33	0.014	51.9	10.9	10.6
0.27	0.015	49.4	11.5	11.2	0.33	0.015	48.5	10.1	9.8
0.27	0.016	46.3	10.8	10.5	0.33	0.016	45.4	9.5	9.2
0.28	0.009	82.1	18.8	18.7	0.34	0.009	80.5	16.5	16.4
0.28	0.010	73.9	17.0	16.8	0.34	0.010	72.4	14.9	14.7
0.28	0.011	67.2	15.4	15.2	0.34	0.011	65.8	13.5	13.3
0.28	0.012	61.6	14.1	13.9	0.34	0.012	60.4	12.4	12.2
0.28	0.013	56.8	13.0	12.8	0.34	0.013	55.7	11.4	11.2
0.28	0.014	52.8	12.1	11.8	0.34	0.014	51.7	10.6	10.4
0.28	0.015	49.3	11.3	11.0	0.34	0.015	48.3	9.9	9.6
0.28	0.016	46.2	10.6	10.3	0.34	0.016	45.3	9.3	9.0
0.29	0.009	81.9	18.4	18.3	0.35	0.009	80.2	16.2	16.0
0.29	0.010	73.7	16.6	16.4	0.35	0.010	72.2	14.5	14.4
0.29	0.011	67.0	15.1	14.9	0.35	0.011	65.6	13.2	13.0
0.29	0.012	61.4	13.8	13.6	0.35	0.012	60.1	12.1	11.9
0.29	0.013	56.7	12.8	12.5	0.35	0.013	55.5	11.2	10.9
0.29	0.014	52.6	11.9	11.6	0.35	0.014	51.5	10.4	10.1
0.29	0.015	49.1	11.1	10.8	0.35	0.015	48.1	9.7	9.4
0.29	0.016	46.0	10.4	10.1	0.35	0.016	45.1	9.1	8.8
0.30	0.009	81.6	18.0	17.9	0.36	0.009	79.9	15.8	15.6
0.30	0.010	73.4	16.2	16.0	0.36	0.010	71.9	14.2	14.0
0.30	0.011	66.8	14.8	14.5	0.36	0.011	65.3	12.9	12.7
0.30	0.012	61.2	13.5	13.3	0.36	0.012	59.9	11.8	11.6
0.30	0.013	56.5	12.5	12.2	0.36	0.013	55.3	10.9	10.7
0.30	0.014	52.5	11.6	11.3	0.36	0.014	51.3	10.2	9.9
0.30	0.015	49.0	10.8	10.5	0.36	0.015	47.9	9.5	9.2
0.30	0.016	45.9	10.2	9.8	0.36	0.016	44.9	8.9	8.6
0.31	0.009	81.3	17.7	17.5	0.37	0.009	79.5	15.4	15.3
0.31	0.010	73.2	15.9	15.7	0.37	0.010	71.6	13.9	13.7
0.31	0.011	66.5	14.4	14.2	0.37	0.011	65.1	12.6	12.4
0.31	0.012	61.0	13.2	13.0	0.37	0.012	59.7	11.6	11.4
0.31	0.013	56.3	12.2	12.0	0.37	0.013	55.1	10.7	10.4
0.31	0.014	52.3	11.4	11.1	0.37	0.014	51.1	9.9	9.7
0.31	0.015	48.8	10.6	10.3	0.37	0.015	47.7	9.3	9.0
0.31	0.016	45.7	9.9	9.6	0.37	0.016	44.7	8.7	8.4

Chart A-1.40 (Continued): Only For Barrels With An Inside Diameter Of 1.40 Inches

(Arbor diameters of from 0.38 Inches thru 0.49 Inches, and spring thicknesses of from 0.009 Thru 0.016 Inches)

For An Arbor Diameter Of	And A Spring Thickness Of	The BEST Spring Length Is	Max Turns Using BEST Length	Max Turns If Spring Is 10" Short Or Long	For An Arbor Diameter Of	And A Spring Thickness Of	The BEST Spring Length Is	Max Turns Using BEST Length	Max Turns If Spring Is 10" Short Or Long
0.38	0.009	79.2	15.1	14.9	0.44	0.009	77.1	13.1	12.9
0.38	0.010	71.3	13.6	13.4	0.44	0.010	69.4	11.8	11.6
0.38	0.011	64.8	12.3	12.1	0.44	0.011	63.1	10.7	10.5
0.38	0.012	59.4	11.3	11.1	0.44	0.012	57.8	9.8	9.6
0.38	0.013	54.8	10.4	10.2	0.44	0.013	53.4	9.1	8.8
0.38	0.014	50.9	9.7	9.4	0.44	0.014	49.5	8.4	8.2
0.38	0.015	47.5	9.1	8.8	0.44	0.015	46.2	7.8	7.6
0.38	0.016	44.6	8.5	8.2	0.44	0.016	43.4	7.4	7.1
0.39	0.009	78.9	14.7	14.6	0.45	0.009	76.7	12.8	12.6
0.39	0.010	71.0	13.3	13.1	0.45	0.010	69.0	11.5	11.3
0.39	0.011	64.5	12.1	11.9	0.45	0.011	62.7	10.4	10.2
0.39	0.012	59.2	11.1	10.8	0.45	0.012	57.5	9.6	9.4
0.39	0.013	54.6	10.2	10.0	0.45	0.013	53.1	8.8	8.6
0.39	0.014	50.7	9.5	9.2	0.45	0.014	49.3	8.2	7.9
0.39	0.015	47.3	8.8	8.6	0.45	0.015	46.0	7.7	7.4
0.39	0.016	44.4	8.3	8.0	0.45	0.016	43.1	7.2	6.9
0.40	0.009	78.5	14.4	14.2	0.46	0.009	76.3	12.4	12.3
0.40	0.010	70.7	13.0	12.8	0.46	0.010	68.7	11.2	11.0
0.40	0.011	64.3	11.8	11.6	0.46	0.011	62.4	10.2	10.0
0.40	0.012	58.9	10.8	10.6	0.46	0.012	57.2	9.3	9.1
0.40	0.013	54.4	10.0	9.7	0.46	0.013	52.8	8.6	8.4
0.40	0.014	50.5	9.3	9.0	0.46	0.014	49.0	8.0	7.7
0.40	0.015	47.1	8.6	8.4	0.46	0.015	45.8	7.5	7.2
0.40	0.016	44.2	8.1	7.8	0.46	0.016	42.9	7.0	6.7
0.41	0.009	78.2	14.1	13.9	0.47	0.009	75.9	12.1	12.0
0.41	0.010	70.4	12.7	12.5	0.47	0.010	68.3	10.9	10.7
0.41	0.011	64.0	11.5	11.3	0.47	0.011	62.1	9.9	9.7
0.41	0.012	58.6	10.5	10.3	0.47	0.012	56.9	9.1	8.9
0.41	0.013	54.1	9.7	9.5	0.47	0.013	52.5	8.4	8.2
0.41	0.014	50.3	9.0	8.8	0.47	0.014	48.8	7.8	7.6
0.41	0.015	46.9	8.4	8.2	0.47	0.015	45.5	7.3	7.0
0.41	0.016	44.0	7.9	7.6	0.47	0.016	42.7	6.8	6.5
0.42	0.009	77.8	13.7	13.6	0.48	0.009	75.5	11.8	11.7
0.42	0.010	70.0	12.4	12.2	0.48	0.010	67.9	10.7	10.5
0.42	0.011	63.7	11.2	11.0	0.48	0.011	61.7	9.7	9.5
0.42	0.012	58.4	10.3	10.1	0.48	0.012	56.6	8.9	8.7
0.42	0.013	53.9	9.5	9.3	0.48	0.013	52.2	8.2	8.0
0.42	0.014	50.0	8.8	8.6	0.48	0.014	48.5	7.6	7.4
0.42	0.015	46.7	8.2	8.0	0.48	0.015	45.3	7.1	6.8
0.42	0.016	43.8	7.7	7.4	0.48	0.016	42.5	6.7	6.4
0.43	0.009	77.5	13.4	13.2	0.49	0.009	75.0	11.5	11.4
0.43	0.010	69.7	12.1	11.9	0.49	0.010	67.5	10.4	10.2
0.43	0.011	63.4	11.0	10.8	0.49	0.011	61.4	9.4	9.2
0.43	0.012	58.1	10.0	9.8	0.49	0.012	56.3	8.7	8.4
0.43	0.013	53.6	9.3	9.0	0.49	0.013	52.0	8.0	7.8
0.43	0.014	49.8	8.6	8.4	0.49	0.014	48.2	7.4	7.2
0.43	0.015	46.5	8.0	7.8	0.49	0.015	45.0	6.9	6.7
0.43	0.016	43.6	7.5	7.2	0.49	0.016	42.2	6.5	6.2

Selecting the Perfect Spring for Your Barreled Spring Clock

Chart A-1.45: Only For Barrels With An Inside Diameter Of 1.45 Inches

(Arbor diameters of from 0.26 Inches thru 0.37 Inches, and spring thicknesses of from 0.009 Thru 0.016 Inches)

For An Arbor Diameter Of	And A Spring Thickness Of	The BEST Spring Length Is	Max Turns Using BEST Length	Max Turns If Spring Is 10" Short Or Long	For An Arbor Diameter Of	And A Spring Thickness Of	The BEST Spring Length Is	Max Turns Using BEST Length	Max Turns If Spring Is 10" Short Or Long
0.26	0.009	88.8	20.7	20.6	0.32	0.009	87.3	18.3	18.2
0.26	0.010	79.9	18.7	18.5	0.32	0.010	78.5	16.5	16.3
0.26	0.011	72.6	17.0	16.8	0.32	0.011	71.4	15.0	14.8
0.26	0.012	66.6	15.6	15.3	0.32	0.012	65.5	13.7	13.5
0.26	0.013	61.5	14.4	14.1	0.32	0.013	60.4	12.7	12.5
0.26	0.014	57.1	13.3	13.1	0.32	0.014	56.1	11.8	11.5
0.26	0.015	53.3	12.4	12.2	0.32	0.015	52.4	11.0	10.7
0.26	0.016	49.9	11.7	11.4	0.32	0.016	49.1	10.3	10.0
0.27	0.009	88.6	20.3	20.2	0.33	0.009	87.0	17.9	17.8
0.27	0.010	79.7	18.3	18.1	0.33	0.010	78.3	16.2	16.0
0.27	0.011	72.5	16.6	16.4	0.33	0.011	71.2	14.7	14.5
0.27	0.012	66.4	15.2	15.0	0.33	0.012	65.2	13.5	13.2
0.27	0.013	61.3	14.1	13.8	0.33	0.013	60.2	12.4	12.2
0.27	0.014	56.9	13.1	12.8	0.33	0.014	55.9	11.5	11.3
0.27	0.015	53.1	12.2	11.9	0.33	0.015	52.2	10.8	10.5
0.27	0.016	49.8	11.4	11.1	0.33	0.016	48.9	10.1	9.8
0.28	0.009	88.3	19.9	19.8	0.34	0.009	86.7	17.6	17.4
0.28	0.010	79.5	17.9	17.7	0.34	0.010	78.0	15.8	15.6
0.28	0.011	72.3	16.3	16.1	0.34	0.011	70.9	14.4	14.2
0.28	0.012	66.2	14.9	14.7	0.34	0.012	65.0	13.2	13.0
0.28	0.013	61.1	13.8	13.6	0.34	0.013	60.0	12.2	11.9
0.28	0.014	56.8	12.8	12.6	0.34	0.014	55.7	11.3	11.0
0.28	0.015	53.0	11.9	11.7	0.34	0.015	52.0	10.5	10.3
0.28	0.016	49.7	11.2	10.9	0.34	0.016	48.8	9.9	9.6
0.29	0.009	88.1	19.5	19.4	0.35	0.009	86.4	17.2	17.0
0.29	0.010	79.3	17.6	17.4	0.35	0.010	77.8	15.5	15.3
0.29	0.011	72.1	16.0	15.8	0.35	0.011	70.7	14.1	13.9
0.29	0.012	66.1	14.6	14.4	0.35	0.012	64.8	12.9	12.7
0.29	0.013	61.0	13.5	13.3	0.35	0.013	59.8	11.9	11.7
0.29	0.014	56.6	12.5	12.3	0.35	0.014	55.5	11.1	10.8
0.29	0.015	52.8	11.7	11.4	0.35	0.015	51.8	10.3	10.1
0.29	0.016	49.5	11.0	10.7	0.35	0.016	48.6	9.7	9.4
0.30	0.009	87.8	19.1	19.0	0.36	0.009	86.1	16.8	16.7
0.30	0.010	79.0	17.2	17.0	0.36	0.010	77.5	15.1	15.0
0.30	0.011	71.8	15.6	15.4	0.36	0.011	70.4	13.8	13.6
0.30	0.012	65.9	14.3	14.1	0.36	0.012	64.6	12.6	12.4
0.30	0.013	60.8	13.2	13.0	0.36	0.013	59.6	11.6	11.4
0.30	0.014	56.5	12.3	12.0	0.36	0.014	55.3	10.8	10.6
0.30	0.015	52.7	11.5	11.2	0.36	0.015	51.7	10.1	9.8
0.30	0.016	49.4	10.8	10.5	0.36	0.016	48.4	9.5	9.2
0.31	0.009	87.5	18.7	18.6	0.37	0.009	85.8	16.5	16.3
0.31	0.010	78.8	16.8	16.7	0.37	0.010	77.2	14.8	14.6
0.31	0.011	71.6	15.3	15.1	0.37	0.011	70.2	13.5	13.3
0.31	0.012	65.7	14.0	13.8	0.37	0.012	64.3	12.3	12.1
0.31	0.013	60.6	13.0	12.7	0.37	0.013	59.4	11.4	11.2
0.31	0.014	56.3	12.0	11.8	0.37	0.014	55.1	10.6	10.3
0.31	0.015	52.5	11.2	11.0	0.37	0.015	51.5	9.9	9.6
0.31	0.016	49.2	10.5	10.2	0.37	0.016	48.2	9.3	9.0

Chart A-1.45 (Continued): Only For Barrels With An Inside Diameter Of 1.45 Inches

(Arbor diameters of from 0.38 Inches thru 0.49 Inches, and spring thicknesses of from 0.009 Thru 0.016 Inches)

For An Arbor Diameter Of	And A Spring Thickness Of	The BEST Spring Length Is	Max Turns Using BEST Length	Max Turns If Spring Is 10" Short Or Long	For An Arbor Diameter Of	And A Spring Thickness Of	The BEST Spring Length Is	Max Turns Using BEST Length	Max Turns If Spring Is 10" Short Or Long
0.38	0.009	85.4	16.1	15.9	0.44	0.009	83.3	14.1	13.9
0.38	0.010	76.9	14.5	14.3	0.44	0.010	75.0	12.6	12.5
0.38	0.011	69.9	13.2	13.0	0.44	0.011	68.1	11.5	11.3
0.38	0.012	64.1	12.1	11.9	0.44	0.012	62.5	10.5	10.3
0.38	0.013	59.1	11.1	10.9	0.44	0.013	57.7	9.7	9.5
0.38	0.014	54.9	10.4	10.1	0.44	0.014	53.5	9.0	8.8
0.38	0.015	51.3	9.7	9.4	0.44	0.015	50.0	8.4	8.2
0.38	0.016	48.1	9.1	8.8	0.44	0.016	46.9	7.9	7.6
0.39	0.009	85.1	15.7	15.6	0.45	0.009	82.9	13.7	13.6
0.39	0.010	76.6	14.2	14.0	0.45	0.010	74.6	12.4	12.2
0.39	0.011	69.6	12.9	12.7	0.45	0.011	67.8	11.2	11.1
0.39	0.012	63.8	11.8	11.6	0.45	0.012	62.2	10.3	10.1
0.39	0.013	58.9	10.9	10.7	0.45	0.013	57.4	9.5	9.3
0.39	0.014	54.7	10.1	9.9	0.45	0.014	53.3	8.8	8.6
0.39	0.015	51.1	9.4	9.2	0.45	0.015	49.7	8.2	8.0
0.39	0.016	47.9	8.9	8.6	0.45	0.016	46.6	7.7	7.5
0.40	0.009	84.8	15.4	15.2	0.46	0.009	82.5	13.4	13.3
0.40	0.010	76.3	13.9	13.7	0.46	0.010	74.3	12.1	11.9
0.40	0.011	69.3	12.6	12.4	0.46	0.011	67.5	11.0	10.8
0.40	0.012	63.6	11.6	11.3	0.46	0.012	61.9	10.1	9.9
0.40	0.013	58.7	10.7	10.4	0.46	0.013	57.1	9.3	9.1
0.40	0.014	54.5	9.9	9.7	0.46	0.014	53.0	8.6	8.4
0.40	0.015	50.9	9.2	9.0	0.46	0.015	49.5	8.0	7.8
0.40	0.016	47.7	8.7	8.4	0.46	0.016	46.4	7.5	7.3
0.41	0.009	84.4	15.1	14.9	0.47	0.009	82.1	13.1	12.9
0.41	0.010	76.0	13.6	13.4	0.47	0.010	73.9	11.8	11.6
0.41	0.011	69.1	12.3	12.1	0.47	0.011	67.2	10.7	10.5
0.41	0.012	63.3	11.3	11.1	0.47	0.012	61.6	9.8	9.6
0.41	0.013	58.4	10.4	10.2	0.47	0.013	56.8	9.1	8.9
0.41	0.014	54.3	9.7	9.4	0.47	0.014	52.8	8.4	8.2
0.41	0.015	50.6	9.0	8.8	0.47	0.015	49.3	7.9	7.6
0.41	0.016	47.5	8.5	8.2	0.47	0.016	46.2	7.4	7.1
0.42	0.009	84.0	14.7	14.6	0.48	0.009	81.7	12.8	12.6
0.42	0.010	75.6	13.2	13.1	0.48	0.010	73.5	11.5	11.3
0.42	0.011	68.8	12.0	11.9	0.48	0.011	66.8	10.5	10.3
0.42	0.012	63.0	11.0	10.8	0.48	0.012	61.3	9.6	9.4
0.42	0.013	58.2	10.2	10.0	0.48	0.013	56.6	8.8	8.6
0.42	0.014	54.0	9.5	9.2	0.48	0.014	52.5	8.2	8.0
0.42	0.015	50.4	8.8	8.6	0.48	0.015	49.0	7.7	7.4
0.42	0.016	47.3	8.3	8.0	0.48	0.016	45.9	7.2	6.9
0.43	0.009	83.7	14.4	14.2	0.49	0.009	81.3	12.5	12.3
0.43	0.010	75.3	12.9	12.8	0.49	0.010	73.1	11.2	11.1
0.43	0.011	68.5	11.8	11.6	0.49	0.011	66.5	10.2	10.0
0.43	0.012	62.8	10.8	10.6	0.49	0.012	60.9	9.4	9.2
0.43	0.013	57.9	10.0	9.7	0.49	0.013	56.3	8.6	8.4
0.43	0.014	53.8	9.2	9.0	0.49	0.014	52.2	8.0	7.8
0.43	0.015	50.2	8.6	8.4	0.49	0.015	48.8	7.5	7.2
0.43	0.016	47.1	8.1	7.8	0.49	0.016	45.7	7.0	6.8

Chart A-1.50: Only For Barrels With An Inside Diameter Of 1.50 Inches

(Arbor diameters of from 0.26 Inches thru 0.37 Inches, and spring thicknesses of from 0.009 Thru 0.016 Inches)

For An Arbor Diameter Of	And A Spring Thickness Of	The BEST Spring Length Is	Max Turns Using BEST Length	Max Turns If Spring Is 10" Short Or Long	For An Arbor Diameter Of	And A Spring Thickness Of	The BEST Spring Length Is	Max Turns Using BEST Length	Max Turns If Spring Is 10" Short Or Long
0.26	0.009	95.2	21.8	21.7	0.32	0.009	93.7	19.4	19.2
0.26	0.010	85.7	19.6	19.5	0.32	0.010	84.3	17.5	17.3
0.26	0.011	77.9	17.9	17.7	0.32	0.011	76.7	15.9	15.7
0.26	0.012	71.4	16.4	16.2	0.32	0.012	70.3	14.5	14.4
0.26	0.013	65.9	15.1	14.9	0.32	0.013	64.9	13.4	13.2
0.26	0.014	61.2	14.0	13.8	0.32	0.014	60.2	12.5	12.2
0.26	0.015	57.1	13.1	12.9	0.32	0.015	56.2	11.6	11.4
0.26	0.016	53.6	12.3	12.0	0.32	0.016	52.7	10.9	10.7
0.27	0.009	95.0	21.4	21.3	0.33	0.009	93.4	19.0	18.9
0.27	0.010	85.5	19.3	19.1	0.33	0.010	84.1	17.1	16.9
0.27	0.011	77.7	17.5	17.3	0.33	0.011	76.4	15.5	15.4
0.27	0.012	71.2	16.1	15.9	0.33	0.012	70.1	14.3	14.1
0.27	0.013	65.8	14.8	14.6	0.33	0.013	64.7	13.2	12.9
0.27	0.014	61.1	13.8	13.5	0.33	0.014	60.1	12.2	12.0
0.27	0.015	57.0	12.8	12.6	0.33	0.015	56.1	11.4	11.2
0.27	0.016	53.4	12.0	11.8	0.33	0.016	52.6	10.7	10.4
0.28	0.009	94.8	21.0	20.9	0.34	0.009	93.1	18.6	18.5
0.28	0.010	85.3	18.9	18.7	0.34	0.010	83.8	16.8	16.6
0.28	0.011	77.5	17.2	17.0	0.34	0.011	76.2	15.2	15.1
0.28	0.012	71.1	15.7	15.6	0.34	0.012	69.8	14.0	13.8
0.28	0.013	65.6	14.5	14.3	0.34	0.013	64.5	12.9	12.7
0.28	0.014	60.9	13.5	13.3	0.34	0.014	59.9	12.0	11.7
0.28	0.015	56.9	12.6	12.4	0.34	0.015	55.9	11.2	10.9
0.28	0.016	53.3	11.8	11.6	0.34	0.016	52.4	10.5	10.2
0.29	0.009	94.5	20.6	20.4	0.35	0.009	92.8	18.2	18.1
0.29	0.010	85.1	18.5	18.4	0.35	0.010	83.5	16.4	16.3
0.29	0.011	77.3	16.8	16.7	0.35	0.011	76.0	14.9	14.7
0.29	0.012	70.9	15.4	15.2	0.35	0.012	69.6	13.7	13.5
0.29	0.013	65.4	14.3	14.0	0.35	0.013	64.3	12.6	12.4
0.29	0.014	60.8	13.2	13.0	0.35	0.014	59.7	11.7	11.5
0.29	0.015	56.7	12.4	12.1	0.35	0.015	55.7	10.9	10.7
0.29	0.016	53.2	11.6	11.3	0.35	0.016	52.2	10.3	10.0
0.30	0.009	94.2	20.2	20.0	0.36	0.009	92.5	17.9	17.7
0.30	0.010	84.8	18.2	18.0	0.36	0.010	83.3	16.1	15.9
0.30	0.011	77.1	16.5	16.3	0.36	0.011	75.7	14.6	14.4
0.30	0.012	70.7	15.1	14.9	0.36	0.012	69.4	13.4	13.2
0.30	0.013	65.2	14.0	13.8	0.36	0.013	64.1	12.4	12.2
0.30	0.014	60.6	13.0	12.8	0.36	0.014	59.5	11.5	11.3
0.30	0.015	56.5	12.1	11.9	0.36	0.015	55.5	10.7	10.5
0.30	0.016	53.0	11.4	11.1	0.36	0.016	52.0	10.0	9.8
0.31	0.009	94.0	19.8	19.6	0.37	0.009	92.2	17.5	17.4
0.31	0.010	84.6	17.8	17.6	0.37	0.010	83.0	15.7	15.6
0.31	0.011	76.9	16.2	16.0	0.37	0.011	75.4	14.3	14.1
0.31	0.012	70.5	14.8	14.6	0.37	0.012	69.2	13.1	12.9
0.31	0.013	65.1	13.7	13.5	0.37	0.013	63.8	12.1	11.9
0.31	0.014	60.4	12.7	12.5	0.37	0.014	59.3	11.2	11.0
0.31	0.015	56.4	11.9	11.6	0.37	0.015	55.3	10.5	10.3
0.31	0.016	52.9	11.1	10.9	0.37	0.016	51.9	9.8	9.6

Chart A-1.50 (Continued): Only For Barrels With An Inside Diameter Of 1.50 Inches

(Arbor diameters of from 0.38 Inches thru 0.49 Inches, and spring thicknesses of from 0.009 Thru 0.016 Inches)

For An Arbor Diameter Of	And A Spring Thickness Of	The BEST Spring Length Is	Max Turns Using BEST Length	Max Turns If Spring Is 10" Short Or Long	For An Arbor Diameter Of	And A Spring Thickness Of	The BEST Spring Length Is	Max Turns Using BEST Length	Max Turns If Spring Is 10" Short Or Long
0.38	0.009	91.9	17.1	17.0	0.44	0.009	89.7	15.0	14.9
0.38	0.010	82.7	15.4	15.3	0.44	0.010	80.8	13.5	13.4
0.38	0.011	75.2	14.0	13.8	0.44	0.011	73.4	12.3	12.1
0.38	0.012	68.9	12.8	12.7	0.44	0.012	67.3	11.3	11.1
0.38	0.013	63.6	11.9	11.7	0.44	0.013	62.1	10.4	10.2
0.38	0.014	59.1	11.0	10.8	0.44	0.014	57.7	9.7	9.5
0.38	0.015	55.1	10.3	10.0	0.44	0.015	53.8	9.0	8.8
0.38	0.016	51.7	9.6	9.4	0.44	0.016	50.5	8.5	8.2
0.39	0.009	91.5	16.8	16.6	0.45	0.009	89.3	14.7	14.6
0.39	0.010	82.4	15.1	14.9	0.45	0.010	80.4	13.2	13.1
0.39	0.011	74.9	13.7	13.6	0.45	0.011	73.1	12.0	11.9
0.39	0.012	68.7	12.6	12.4	0.45	0.012	67.0	11.0	10.9
0.39	0.013	63.4	11.6	11.4	0.45	0.013	61.9	10.2	10.0
0.39	0.014	58.8	10.8	10.6	0.45	0.014	57.4	9.5	9.2
0.39	0.015	54.9	10.1	9.8	0.45	0.015	53.6	8.8	8.6
0.39	0.016	51.5	9.4	9.2	0.45	0.016	50.3	8.3	8.0
0.40	0.009	91.2	16.4	16.3	0.46	0.009	88.9	14.4	14.2
0.40	0.010	82.1	14.8	14.6	0.46	0.010	80.0	12.9	12.8
0.40	0.011	74.6	13.4	13.3	0.46	0.011	72.8	11.8	11.6
0.40	0.012	68.4	12.3	12.1	0.46	0.012	66.7	10.8	10.6
0.40	0.013	63.1	11.4	11.2	0.46	0.013	61.6	10.0	9.8
0.40	0.014	58.6	10.6	10.3	0.46	0.014	57.2	9.2	9.0
0.40	0.015	54.7	9.8	9.6	0.46	0.015	53.4	8.6	8.4
0.40	0.016	51.3	9.2	9.0	0.46	0.016	50.0	8.1	7.8
0.41	0.009	90.8	16.1	15.9	0.47	0.009	88.5	14.1	13.9
0.41	0.010	81.8	14.5	14.3	0.47	0.010	79.7	12.7	12.5
0.41	0.011	74.3	13.1	13.0	0.47	0.011	72.4	11.5	11.3
0.41	0.012	68.1	12.0	11.9	0.47	0.012	66.4	10.5	10.4
0.41	0.013	62.9	11.1	10.9	0.47	0.013	61.3	9.7	9.5
0.41	0.014	58.4	10.3	10.1	0.47	0.014	56.9	9.0	8.8
0.41	0.015	54.5	9.6	9.4	0.47	0.015	53.1	8.4	8.2
0.41	0.016	51.1	9.0	8.8	0.47	0.016	49.8	7.9	7.7
0.42	0.009	90.5	15.7	15.6	0.48	0.009	88.1	13.7	13.6
0.42	0.010	81.4	14.1	14.0	0.48	0.010	79.3	12.4	12.2
0.42	0.011	74.0	12.9	12.7	0.48	0.011	72.1	11.2	11.1
0.42	0.012	67.9	11.8	11.6	0.48	0.012	66.1	10.3	10.1
0.42	0.013	62.6	10.9	10.7	0.48	0.013	61.0	9.5	9.3
0.42	0.014	58.2	10.1	9.9	0.48	0.014	56.6	8.8	8.6
0.42	0.015	54.3	9.4	9.2	0.48	0.015	52.9	8.2	8.0
0.42	0.016	50.9	8.8	8.6	0.48	0.016	49.6	7.7	7.5
0.43	0.009	90.1	15.4	15.2	0.49	0.009	87.7	13.4	13.3
0.43	0.010	81.1	13.8	13.7	0.49	0.010	78.9	12.1	11.9
0.43	0.011	73.7	12.6	12.4	0.49	0.011	71.8	11.0	10.8
0.43	0.012	67.6	11.5	11.3	0.49	0.012	65.8	10.1	9.9
0.43	0.013	62.4	10.6	10.4	0.49	0.013	60.7	9.3	9.1
0.43	0.014	57.9	9.9	9.7	0.49	0.014	56.4	8.6	8.4
0.43	0.015	54.1	9.2	9.0	0.49	0.015	52.6	8.1	7.8
0.43	0.016	50.7	8.6	8.4	0.49	0.016	49.3	7.6	7.3

Chart A-1.55: Only For Barrels With An Inside Diameter Of 1.55 Inches

(Arbor diameters of from 0.26 Inches thru 0.37 Inches, and spring thicknesses of from 0.010 Thru 0.017 Inches)

For An Arbor Diameter Of	And A Spring Thickness Of	The BEST Spring Length Is	Max Turns Using BEST Length	Max Turns If Spring Is 10" Short Or Long	For An Arbor Diameter Of	And A Spring Thickness Of	The BEST Spring Length Is	Max Turns Using BEST Length	Max Turns If Spring Is 10" Short Or Long
0.26	0.010	91.7	20.6	20.5	0.32	0.010	90.3	18.4	18.3
0.26	0.011	83.4	18.8	18.6	0.32	0.011	82.1	16.7	16.6
0.26	0.012	76.4	17.2	17.0	0.32	0.012	75.3	15.3	15.2
0.26	0.013	70.5	15.9	15.7	0.32	0.013	69.5	14.2	14.0
0.26	0.014	65.5	14.7	14.5	0.32	0.014	64.5	13.2	12.9
0.26	0.015	61.1	13.8	13.5	0.32	0.015	60.2	12.3	12.1
0.26	0.016	57.3	12.9	12.7	0.32	0.016	56.5	11.5	11.3
0.26	0.017	53.9	12.1	11.9	0.32	0.017	53.1	10.8	10.6
0.27	0.010	91.5	20.3	20.1	0.33	0.010	90.1	18.1	17.9
0.27	0.011	83.2	18.4	18.2	0.33	0.011	81.9	16.4	16.3
0.27	0.012	76.2	16.9	16.7	0.33	0.012	75.1	15.0	14.9
0.27	0.013	70.4	15.6	15.4	0.33	0.013	69.3	13.9	13.7
0.27	0.014	65.3	14.5	14.3	0.33	0.014	64.3	12.9	12.7
0.27	0.015	61.0	13.5	13.3	0.33	0.015	60.0	12.0	11.8
0.27	0.016	57.2	12.7	12.4	0.33	0.016	56.3	11.3	11.1
0.27	0.017	53.8	11.9	11.7	0.33	0.017	53.0	10.6	10.4
0.28	0.010	91.3	19.9	19.7	0.34	0.010	89.8	17.7	17.6
0.28	0.011	83.0	18.1	17.9	0.34	0.011	81.6	16.1	15.9
0.28	0.012	76.1	16.6	16.4	0.34	0.012	74.8	14.8	14.6
0.28	0.013	70.2	15.3	15.1	0.34	0.013	69.1	13.6	13.4
0.28	0.014	65.2	14.2	14.0	0.34	0.014	64.1	12.6	12.4
0.28	0.015	60.8	13.3	13.0	0.34	0.015	59.9	11.8	11.6
0.28	0.016	57.0	12.4	12.2	0.34	0.016	56.1	11.1	10.8
0.28	0.017	53.7	11.7	11.4	0.34	0.017	52.8	10.4	10.2
0.29	0.010	91.0	19.5	19.4	0.35	0.010	89.5	17.4	17.2
0.29	0.011	82.8	17.7	17.6	0.35	0.011	81.4	15.8	15.6
0.29	0.012	75.9	16.3	16.1	0.35	0.012	74.6	14.5	14.3
0.29	0.013	70.0	15.0	14.8	0.35	0.013	68.9	13.4	13.2
0.29	0.014	65.0	13.9	13.7	0.35	0.014	64.0	12.4	12.2
0.29	0.015	60.7	13.0	12.8	0.35	0.015	59.7	11.6	11.4
0.29	0.016	56.9	12.2	12.0	0.35	0.016	56.0	10.9	10.6
0.29	0.017	53.6	11.5	11.2	0.35	0.017	52.7	10.2	10.0
0.30	0.010	90.8	19.1	19.0	0.36	0.010	89.3	17.0	16.9
0.30	0.011	82.6	17.4	17.2	0.36	0.011	81.1	15.5	15.3
0.30	0.012	75.7	15.9	15.8	0.36	0.012	74.4	14.2	14.0
0.30	0.013	69.9	14.7	14.5	0.36	0.013	68.7	13.1	12.9
0.30	0.014	64.9	13.7	13.5	0.36	0.014	63.8	12.2	12.0
0.30	0.015	60.5	12.8	12.5	0.36	0.015	59.5	11.3	11.1
0.30	0.016	56.8	12.0	11.7	0.36	0.016	55.8	10.6	10.4
0.30	0.017	53.4	11.3	11.0	0.36	0.017	52.5	10.0	9.8
0.31	0.010	90.6	18.8	18.6	0.37	0.010	89.0	16.7	16.5
0.31	0.011	82.3	17.1	16.9	0.37	0.011	80.9	15.2	15.0
0.31	0.012	75.5	15.6	15.5	0.37	0.012	74.1	13.9	13.7
0.31	0.013	69.7	14.4	14.3	0.37	0.013	68.4	12.8	12.6
0.31	0.014	64.7	13.4	13.2	0.37	0.014	63.5	11.9	11.7
0.31	0.015	60.4	12.5	12.3	0.37	0.015	59.3	11.1	10.9
0.31	0.016	56.6	11.7	11.5	0.37	0.016	55.6	10.4	10.2
0.31	0.017	53.3	11.0	10.8	0.37	0.017	52.3	9.8	9.6

Chart A-1.55 (Continued): Only For Barrels With An Inside Diameter Of 1.55 Inches

(Arbor diameters of from 0.38 Inches thru 0.49 Inches, and spring thicknesses of from 0.010 Thru 0.017 Inches)

For An Arbor Diameter Of	And A Spring Thickness Of	The BEST Spring Length Is	Max Turns Using BEST Length	Max Turns If Spring Is 10" Short Or Long	For An Arbor Diameter Of	And A Spring Thickness Of	The BEST Spring Length Is	Max Turns Using BEST Length	Max Turns If Spring Is 10" Short Or Long
0.38	0.010	88.7	16.3	16.2	0.44	0.010	86.7	14.4	14.3
0.38	0.011	80.6	14.9	14.7	0.44	0.011	78.9	13.1	13.0
0.38	0.012	73.9	13.6	13.5	0.44	0.012	72.3	12.0	11.9
0.38	0.013	68.2	12.6	12.4	0.44	0.013	66.7	11.1	10.9
0.38	0.014	63.3	11.7	11.5	0.44	0.014	62.0	10.3	10.1
0.38	0.015	59.1	10.9	10.7	0.44	0.015	57.8	9.6	9.4
0.38	0.016	55.4	10.2	10.0	0.44	0.016	54.2	9.0	8.8
0.38	0.017	52.2	9.6	9.4	0.44	0.017	51.0	8.5	8.3
0.39	0.010	88.4	16.0	15.9	0.45	0.010	86.4	14.1	14.0
0.39	0.011	80.3	14.6	14.4	0.45	0.011	78.5	12.8	12.7
0.39	0.012	73.6	13.3	13.2	0.45	0.012	72.0	11.8	11.6
0.39	0.013	68.0	12.3	12.1	0.45	0.013	66.5	10.9	10.7
0.39	0.014	63.1	11.4	11.2	0.45	0.014	61.7	10.1	9.9
0.39	0.015	58.9	10.7	10.5	0.45	0.015	57.6	9.4	9.2
0.39	0.016	55.2	10.0	9.8	0.45	0.016	54.0	8.8	8.6
0.39	0.017	52.0	9.4	9.2	0.45	0.017	50.8	8.3	8.1
0.40	0.010	88.1	15.7	15.6	0.46	0.010	86.0	13.8	13.7
0.40	0.011	80.1	14.3	14.1	0.46	0.011	78.2	12.6	12.4
0.40	0.012	73.4	13.1	12.9	0.46	0.012	71.7	11.5	11.4
0.40	0.013	67.7	12.1	11.9	0.46	0.013	66.2	10.6	10.5
0.40	0.014	62.9	11.2	11.0	0.46	0.014	61.5	9.9	9.7
0.40	0.015	58.7	10.5	10.3	0.46	0.015	57.4	9.2	9.0
0.40	0.016	55.0	9.8	9.6	0.46	0.016	53.8	8.6	8.4
0.40	0.017	51.8	9.2	9.0	0.46	0.017	50.6	8.1	7.9
0.41	0.010	87.7	15.4	15.2	0.47	0.010	85.7	13.5	13.4
0.41	0.011	79.8	14.0	13.8	0.47	0.011	77.9	12.3	12.2
0.41	0.012	73.1	12.8	12.6	0.47	0.012	71.4	11.3	11.1
0.41	0.013	67.5	11.8	11.6	0.47	0.013	65.9	10.4	10.2
0.41	0.014	62.7	11.0	10.8	0.47	0.014	61.2	9.7	9.5
0.41	0.015	58.5	10.2	10.0	0.47	0.015	57.1	9.0	8.8
0.41	0.016	54.8	9.6	9.4	0.47	0.016	53.5	8.5	8.2
0.41	0.017	51.6	9.0	8.8	0.47	0.017	50.4	8.0	7.7
0.42	0.010	87.4	15.1	14.9	0.48	0.010	85.3	13.2	13.1
0.42	0.011	79.5	13.7	13.5	0.48	0.011	77.5	12.0	11.9
0.42	0.012	72.8	12.5	12.4	0.48	0.012	71.1	11.0	10.9
0.42	0.013	67.2	11.6	11.4	0.48	0.013	65.6	10.2	10.0
0.42	0.014	62.4	10.8	10.6	0.48	0.014	60.9	9.5	9.3
0.42	0.015	58.3	10.0	9.8	0.48	0.015	56.9	8.8	8.6
0.42	0.016	54.6	9.4	9.2	0.48	0.016	53.3	8.3	8.1
0.42	0.017	51.4	8.9	8.6	0.48	0.017	50.2	7.8	7.6
0.43	0.010	87.1	14.7	14.6	0.49	0.010	84.9	12.9	12.8
0.43	0.011	79.2	13.4	13.2	0.49	0.011	77.2	11.8	11.6
0.43	0.012	72.6	12.3	12.1	0.49	0.012	70.8	10.8	10.6
0.43	0.013	67.0	11.3	11.2	0.49	0.013	65.3	10.0	9.8
0.43	0.014	62.2	10.5	10.3	0.49	0.014	60.7	9.2	9.1
0.43	0.015	58.1	9.8	9.6	0.49	0.015	56.6	8.6	8.4
0.43	0.016	54.4	9.2	9.0	0.49	0.016	53.1	8.1	7.9
0.43	0.017	51.2	8.7	8.4	0.49	0.017	50.0	7.6	7.4

Chart A-1.60: Only For Barrels With An Inside Diameter Of 1.60 Inches

(Arbor diameters of from 0.26 Inches thru 0.37 Inches, and spring thicknesses of from 0.010 Thru 0.017 Inches)

For An Arbor Diameter Of	And A Spring Thickness Of	The BEST Spring Length Is	Max Turns Using BEST Length	Max Turns If Spring Is 10" Short Or Long	For An Arbor Diameter Of	And A Spring Thickness Of	The BEST Spring Length Is	Max Turns Using BEST Length	Max Turns If Spring Is 10" Short Or Long
0.26	0.010	97.9	21.6	21.5	0.32	0.010	96.5	19.4	19.2
0.26	0.011	89.0	19.7	19.5	0.32	0.011	87.7	17.6	17.5
0.26	0.012	81.6	18.0	17.9	0.32	0.012	80.4	16.1	16.0
0.26	0.013	75.3	16.6	16.5	0.32	0.013	74.2	14.9	14.7
0.26	0.014	69.9	15.4	15.3	0.32	0.014	68.9	13.8	13.7
0.26	0.015	65.3	14.4	14.2	0.32	0.015	64.3	12.9	12.7
0.26	0.016	61.2	13.5	13.3	0.32	0.016	60.3	12.1	11.9
0.26	0.017	57.6	12.7	12.5	0.32	0.017	56.8	11.4	11.2
0.27	0.010	97.7	21.2	21.1	0.33	0.010	96.3	19.0	18.9
0.27	0.011	88.8	19.3	19.2	0.33	0.011	87.5	17.3	17.1
0.27	0.012	81.4	17.7	17.5	0.33	0.012	80.2	15.8	15.7
0.27	0.013	75.1	16.3	16.2	0.33	0.013	74.0	14.6	14.5
0.27	0.014	69.8	15.2	15.0	0.33	0.014	68.8	13.6	13.4
0.27	0.015	65.1	14.2	14.0	0.33	0.015	64.2	12.7	12.5
0.27	0.016	61.0	13.3	13.1	0.33	0.016	60.2	11.9	11.7
0.27	0.017	57.5	12.5	12.3	0.33	0.017	56.6	11.2	11.0
0.28	0.010	97.5	20.9	20.7	0.34	0.010	96.0	18.7	18.5
0.28	0.011	88.6	19.0	18.8	0.34	0.011	87.3	17.0	16.8
0.28	0.012	81.2	17.4	17.2	0.34	0.012	80.0	15.6	15.4
0.28	0.013	75.0	16.0	15.9	0.34	0.013	73.8	14.4	14.2
0.28	0.014	69.6	14.9	14.7	0.34	0.014	68.6	13.3	13.1
0.28	0.015	65.0	13.9	13.7	0.34	0.015	64.0	12.4	12.2
0.28	0.016	60.9	13.0	12.8	0.34	0.016	60.0	11.7	11.5
0.28	0.017	57.3	12.3	12.0	0.34	0.017	56.5	11.0	10.8
0.29	0.010	97.2	20.5	20.3	0.35	0.010	95.7	18.3	18.2
0.29	0.011	88.4	18.6	18.5	0.35	0.011	87.0	16.6	16.5
0.29	0.012	81.0	17.1	16.9	0.35	0.012	79.8	15.3	15.1
0.29	0.013	74.8	15.8	15.6	0.35	0.013	73.6	14.1	13.9
0.29	0.014	69.4	14.6	14.4	0.35	0.014	68.4	13.1	12.9
0.29	0.015	64.8	13.7	13.5	0.35	0.015	63.8	12.2	12.0
0.29	0.016	60.8	12.8	12.6	0.35	0.016	59.8	11.4	11.2
0.29	0.017	57.2	12.0	11.8	0.35	0.017	56.3	10.8	10.5
0.30	0.010	97.0	20.1	20.0	0.36	0.010	95.4	18.0	17.8
0.30	0.011	88.2	18.3	18.1	0.36	0.011	86.8	16.3	16.2
0.30	0.012	80.8	16.8	16.6	0.36	0.012	79.5	15.0	14.8
0.30	0.013	74.6	15.5	15.3	0.36	0.013	73.4	13.8	13.6
0.30	0.014	69.3	14.4	14.2	0.36	0.014	68.2	12.8	12.6
0.30	0.015	64.7	13.4	13.2	0.36	0.015	63.6	12.0	11.8
0.30	0.016	60.6	12.6	12.4	0.36	0.016	59.7	11.2	11.0
0.30	0.017	57.1	11.8	11.6	0.36	0.017	56.1	10.6	10.3
0.31	0.010	96.8	19.7	19.6	0.37	0.010	95.2	17.6	17.5
0.31	0.011	88.0	17.9	17.8	0.37	0.011	86.5	16.0	15.9
0.31	0.012	80.6	16.5	16.3	0.37	0.012	79.3	14.7	14.5
0.31	0.013	74.4	15.2	15.0	0.37	0.013	73.2	13.6	13.4
0.31	0.014	69.1	14.1	13.9	0.37	0.014	68.0	12.6	12.4
0.31	0.015	64.5	13.2	13.0	0.37	0.015	63.4	11.7	11.6
0.31	0.016	60.5	12.3	12.1	0.37	0.016	59.5	11.0	10.8
0.31	0.017	56.9	11.6	11.4	0.37	0.017	56.0	10.4	10.1

Chart A-1.60 (Continued): Only For Barrels With An Inside Diameter Of 1.60 Inches

(Arbor diameters of from 0.38 Inches thru 0.49 Inches, and spring thicknesses of from 0.010 Thru 0.017 Inches)

For An Arbor Diameter Of	And A Spring Thickness Of	The BEST Spring Length Is	Max Turns Using BEST Length	Max Turns If Spring Is 10" Short Or Long	For An Arbor Diameter Of	And A Spring Thickness Of	The BEST Spring Length Is	Max Turns Using BEST Length	Max Turns If Spring Is 10" Short Or Long
0.38	0.010	94.9	17.3	17.2	0.44	0.010	92.9	15.3	15.2
0.38	0.011	86.2	15.7	15.6	0.44	0.011	84.5	13.9	13.8
0.38	0.012	79.1	14.4	14.2	0.44	0.012	77.4	12.8	12.6
0.38	0.013	73.0	13.3	13.1	0.44	0.013	71.5	11.8	11.6
0.38	0.014	67.8	12.3	12.2	0.44	0.014	66.4	11.0	10.8
0.38	0.015	63.2	11.5	11.3	0.44	0.015	62.0	10.2	10.0
0.38	0.016	59.3	10.8	10.6	0.44	0.016	58.1	9.6	9.4
0.38	0.017	55.8	10.2	9.9	0.44	0.017	54.7	9.0	8.8
0.39	0.010	94.6	16.9	16.8	0.45	0.010	92.6	15.0	14.9
0.39	0.011	86.0	15.4	15.3	0.45	0.011	84.2	13.7	13.5
0.39	0.012	78.8	14.1	14.0	0.45	0.012	77.1	12.5	12.4
0.39	0.013	72.7	13.0	12.9	0.45	0.013	71.2	11.6	11.4
0.39	0.014	67.5	12.1	11.9	0.45	0.014	66.1	10.7	10.6
0.39	0.015	63.0	11.3	11.1	0.45	0.015	61.7	10.0	9.8
0.39	0.016	59.1	10.6	10.4	0.45	0.016	57.9	9.4	9.2
0.39	0.017	55.6	10.0	9.8	0.45	0.017	54.5	8.8	8.6
0.40	0.010	94.2	16.6	16.5	0.46	0.010	92.2	14.7	14.6
0.40	0.011	85.7	15.1	15.0	0.46	0.011	83.8	13.4	13.2
0.40	0.012	78.5	13.8	13.7	0.46	0.012	76.9	12.3	12.1
0.40	0.013	72.5	12.8	12.6	0.46	0.013	70.9	11.3	11.2
0.40	0.014	67.3	11.9	11.7	0.46	0.014	65.9	10.5	10.3
0.40	0.015	62.8	11.1	10.9	0.46	0.015	61.5	9.8	9.6
0.40	0.016	58.9	10.4	10.2	0.46	0.016	57.6	9.2	9.0
0.40	0.017	55.4	9.8	9.6	0.46	0.017	54.2	8.7	8.4
0.41	0.010	93.9	16.3	16.2	0.47	0.010	91.9	14.4	14.3
0.41	0.011	85.4	14.8	14.7	0.47	0.011	83.5	13.1	13.0
0.41	0.012	78.3	13.6	13.4	0.47	0.012	76.5	12.0	11.9
0.41	0.013	72.3	12.5	12.4	0.47	0.013	70.7	11.1	10.9
0.41	0.014	67.1	11.6	11.5	0.47	0.014	65.6	10.3	10.1
0.41	0.015	62.6	10.9	10.7	0.47	0.015	61.2	9.6	9.4
0.41	0.016	58.7	10.2	10.0	0.47	0.016	57.4	9.0	8.8
0.41	0.017	55.3	9.6	9.4	0.47	0.017	54.0	8.5	8.3
0.42	0.010	93.6	16.0	15.8	0.48	0.010	91.5	14.1	14.0
0.42	0.011	85.1	14.5	14.4	0.48	0.011	83.2	12.8	12.7
0.42	0.012	78.0	13.3	13.2	0.48	0.012	76.2	11.8	11.6
0.42	0.013	72.0	12.3	12.1	0.48	0.013	70.4	10.9	10.7
0.42	0.014	66.9	11.4	11.2	0.48	0.014	65.3	10.1	9.9
0.42	0.015	62.4	10.6	10.5	0.48	0.015	61.0	9.4	9.2
0.42	0.016	58.5	10.0	9.8	0.48	0.016	57.2	8.8	8.6
0.42	0.017	55.1	9.4	9.2	0.48	0.017	53.8	8.3	8.1
0.43	0.010	93.3	15.7	15.5	0.49	0.010	91.1	13.8	13.7
0.43	0.011	84.8	14.2	14.1	0.49	0.011	82.8	12.6	12.4
0.43	0.012	77.7	13.0	12.9	0.49	0.012	75.9	11.5	11.4
0.43	0.013	71.7	12.0	11.9	0.49	0.013	70.1	10.6	10.5
0.43	0.014	66.6	11.2	11.0	0.49	0.014	65.1	9.9	9.7
0.43	0.015	62.2	10.4	10.2	0.49	0.015	60.7	9.2	9.0
0.43	0.016	58.3	9.8	9.6	0.49	0.016	56.9	8.6	8.4
0.43	0.017	54.9	9.2	9.0	0.49	0.017	53.6	8.1	7.9

Chart A-1.65: Only For Barrels With An Inside Diameter Of 1.65 Inches

(Arbor diameters of from 0.26 Inches thru 0.37 Inches, and spring thicknesses of from 0.010 Thru 0.017 Inches)

For An Arbor Diameter Of	And A Spring Thickness Of	The BEST Spring Length Is	Max Turns Using BEST Length	Max Turns If Spring Is 10" Short Or Long	For An Arbor Diameter Of	And A Spring Thickness Of	The BEST Spring Length Is	Max Turns Using BEST Length	Max Turns If Spring Is 10" Short Or Long
0.26	0.010	104.3	22.6	22.5	0.32	0.010	102.9	20.3	20.2
0.26	0.011	94.8	20.6	20.4	0.32	0.011	93.5	18.5	18.4
0.26	0.012	86.9	18.8	18.7	0.32	0.012	85.7	17.0	16.8
0.26	0.013	80.2	17.4	17.2	0.32	0.013	79.1	15.7	15.5
0.26	0.014	74.5	16.2	16.0	0.32	0.014	73.5	14.5	14.4
0.26	0.015	69.5	15.1	14.9	0.32	0.015	68.6	13.6	13.4
0.26	0.016	65.2	14.1	13.9	0.32	0.016	64.3	12.7	12.5
0.26	0.017	61.3	13.3	13.1	0.32	0.017	60.5	12.0	11.8
0.27	0.010	104.0	22.2	22.1	0.33	0.010	102.6	19.98	19.86
0.27	0.011	94.6	20.2	20.1	0.33	0.011	93.3	18.17	18.03
0.27	0.012	86.7	18.5	18.4	0.33	0.012	85.5	16.65	16.51
0.27	0.013	80.0	17.1	16.9	0.33	0.013	79.0	15.37	15.21
0.27	0.014	74.3	15.9	15.7	0.33	0.014	73.3	14.27	14.10
0.27	0.015	69.4	14.8	14.6	0.33	0.015	68.4	13.32	13.14
0.27	0.016	65.0	13.9	13.7	0.33	0.016	64.1	12.49	12.30
0.27	0.017	61.2	13.1	12.9	0.33	0.017	60.4	11.75	11.55
0.28	0.010	103.8	21.8	21.7	0.34	0.010	102.4	19.6	19.5
0.28	0.011	94.4	19.9	19.7	0.34	0.011	93.1	17.8	17.7
0.28	0.012	86.5	18.2	18.1	0.34	0.012	85.3	16.4	16.2
0.28	0.013	79.9	16.8	16.6	0.34	0.013	78.7	15.1	14.9
0.28	0.014	74.2	15.6	15.4	0.34	0.014	73.1	14.0	13.8
0.28	0.015	69.2	14.6	14.4	0.34	0.015	68.2	13.1	12.9
0.28	0.016	64.9	13.7	13.5	0.34	0.016	64.0	12.3	12.1
0.28	0.017	61.1	12.8	12.6	0.34	0.017	60.2	11.5	11.3
0.29	0.010	103.6	21.5	21.3	0.35	0.010	102.1	19.3	19.1
0.29	0.011	94.2	19.5	19.4	0.35	0.011	92.8	17.5	17.4
0.29	0.012	86.3	17.9	17.7	0.35	0.012	85.1	16.1	15.9
0.29	0.013	79.7	16.5	16.3	0.35	0.013	78.5	14.8	14.7
0.29	0.014	74.0	15.3	15.2	0.35	0.014	72.9	13.8	13.6
0.29	0.015	69.1	14.3	14.1	0.35	0.015	68.1	12.8	12.7
0.29	0.016	64.8	13.4	13.2	0.35	0.016	63.8	12.0	11.9
0.29	0.017	60.9	12.6	12.4	0.35	0.017	60.1	11.3	11.1
0.30	0.010	103.4	21.1	21.0	0.36	0.010	101.8	18.9	18.8
0.30	0.011	94.0	19.2	19.0	0.36	0.011	92.6	17.2	17.1
0.30	0.012	86.1	17.6	17.4	0.36	0.012	84.9	15.8	15.6
0.30	0.013	79.5	16.2	16.1	0.36	0.013	78.3	14.6	14.4
0.30	0.014	73.8	15.1	14.9	0.36	0.014	72.7	13.5	13.3
0.30	0.015	68.9	14.1	13.9	0.36	0.015	67.9	12.6	12.4
0.30	0.016	64.6	13.2	13.0	0.36	0.016	63.6	11.8	11.6
0.30	0.017	60.8	12.4	12.2	0.36	0.017	59.9	11.1	10.9
0.31	0.010	103.1	20.7	20.6	0.37	0.010	101.5	18.6	18.5
0.31	0.011	93.8	18.8	18.7	0.37	0.011	92.3	16.9	16.8
0.31	0.012	85.9	17.3	17.1	0.37	0.012	84.6	15.5	15.3
0.31	0.013	79.3	15.9	15.8	0.37	0.013	78.1	14.3	14.1
0.31	0.014	73.7	14.8	14.6	0.37	0.014	72.5	13.3	13.1
0.31	0.015	68.8	13.8	13.6	0.37	0.015	67.7	12.4	12.2
0.31	0.016	64.5	12.9	12.8	0.37	0.016	63.5	11.6	11.4
0.31	0.017	60.7	12.2	12.0	0.37	0.017	59.7	10.9	10.7

Chart A-1.65 (Continued): Only For Barrels With An Inside Diameter Of 1.65 Inches

(Arbor diameters of from 0.38 Inches thru 0.49 Inches, and spring thicknesses of from 0.010 Thru 0.017 Inches)

For An Arbor Diameter Of	And A Spring Thickness Of	The BEST Spring Length Is	Max Turns Using BEST Length	Max Turns If Spring Is 10" Short Or Long	For An Arbor Diameter Of	And A Spring Thickness Of	The BEST Spring Length Is	Max Turns Using BEST Length	Max Turns If Spring Is 10" Short Or Long
0.38	0.010	101.2	18.2	18.1	0.44	0.010	99.3	16.2	16.1
0.38	0.011	92.0	16.6	16.4	0.44	0.011	90.3	14.8	14.6
0.38	0.012	84.4	15.2	15.0	0.44	0.012	82.8	13.5	13.4
0.38	0.013	77.9	14.0	13.9	0.44	0.013	76.4	12.5	12.3
0.38	0.014	72.3	13.0	12.9	0.44	0.014	70.9	11.6	11.4
0.38	0.015	67.5	12.2	12.0	0.44	0.015	66.2	10.8	10.7
0.38	0.016	63.3	11.4	11.2	0.44	0.016	62.1	10.2	10.0
0.38	0.017	59.6	10.7	10.5	0.44	0.017	58.4	9.6	9.4
0.39	0.010	100.9	17.9	17.8	0.45	0.010	99.0	15.9	15.8
0.39	0.011	91.8	16.3	16.1	0.45	0.011	90.0	14.5	14.4
0.39	0.012	84.1	14.9	14.8	0.45	0.012	82.5	13.3	13.1
0.39	0.013	77.6	13.8	13.6	0.45	0.013	76.1	12.3	12.1
0.39	0.014	72.1	12.8	12.6	0.45	0.014	70.7	11.4	11.2
0.39	0.015	67.3	11.9	11.7	0.45	0.015	66.0	10.6	10.4
0.39	0.016	63.1	11.2	11.0	0.45	0.016	61.9	10.0	9.8
0.39	0.017	59.4	10.5	10.3	0.45	0.017	58.2	9.4	9.2
0.40	0.010	100.6	17.6	17.4	0.46	0.010	98.6	15.6	15.5
0.40	0.011	91.5	16.0	15.8	0.46	0.011	89.6	14.2	14.1
0.40	0.012	83.9	14.6	14.5	0.46	0.012	82.2	13.0	12.9
0.40	0.013	77.4	13.5	13.3	0.46	0.013	75.8	12.0	11.9
0.40	0.014	71.9	12.5	12.4	0.46	0.014	70.4	11.2	11.0
0.40	0.015	67.1	11.7	11.5	0.46	0.015	65.7	10.4	10.2
0.40	0.016	62.9	11.0	10.8	0.46	0.016	61.6	9.8	9.6
0.40	0.017	59.2	10.3	10.1	0.46	0.017	58.0	9.2	9.0
0.41	0.010	100.3	17.22	17	0.47	0.010	98.2	15.3	15.2
0.41	0.011	91.2	15.66	16	0.47	0.011	89.3	13.9	13.8
0.41	0.012	83.6	14.35	14	0.47	0.012	81.9	12.8	12.6
0.41	0.013	77.2	13.25	13	0.47	0.013	75.6	11.8	11.6
0.41	0.014	71.7	12.30	12	0.47	0.014	70.2	10.9	10.8
0.41	0.015	66.9	11.48	11	0.47	0.015	65.5	10.2	10.0
0.41	0.016	62.7	10.76	11	0.47	0.016	61.4	9.6	9.4
0.41	0.017	59.0	10.13	10	0.47	0.017	57.8	9.0	8.8
0.42	0.010	100.0	16.9	16.8	0.48	0.010	97.9	15.0	14.9
0.42	0.011	90.9	15.4	15.2	0.48	0.011	89.0	13.6	13.5
0.42	0.012	83.3	14.1	13.9	0.48	0.012	81.6	12.5	12.4
0.42	0.013	76.9	13.0	12.8	0.48	0.013	75.3	11.5	11.4
0.42	0.014	71.4	12.1	11.9	0.48	0.014	69.9	10.7	10.6
0.42	0.015	66.7	11.3	11.1	0.48	0.015	65.2	10.0	9.8
0.42	0.016	62.5	10.6	10.4	0.48	0.016	61.2	9.4	9.2
0.42	0.017	58.8	9.9	9.7	0.48	0.017	57.6	8.8	8.6
0.43	0.010	99.7	16.6	16.5	0.49	0.010	97.5	14.71	14.60
0.43	0.011	90.6	15.1	14.9	0.49	0.011	88.6	13.37	13.25
0.43	0.012	83.0	13.8	13.7	0.49	0.012	81.2	12.26	12.12
0.43	0.013	76.7	12.7	12.6	0.49	0.013	75.0	11.31	11.17
0.43	0.014	71.2	11.8	11.7	0.49	0.014	69.6	10.51	10.35
0.43	0.015	66.4	11.0	10.9	0.49	0.015	65.0	9.81	9.64
0.43	0.016	62.3	10.4	10.2	0.49	0.016	60.9	9.19	9.01
0.43	0.017	58.6	9.7	9.5	0.49	0.017	57.3	8.65	8.46

Selecting the Perfect Spring for Your Barreled Spring Clock

Chart A-1.70: Only For Barrels With An Inside Diameter Of 1.70 Inches

(Arbor diameters of from 0.26 Inches thru 0.37 Inches, and spring thicknesses of from 0.010 Thru 0.017 Inches)

For An Arbor Diameter Of	And A Spring Thickness Of	The BEST Spring Length Is	Max Turns Using BEST Length	Max Turns If Spring Is 10" Short Or Long	For An Arbor Diameter Of	And A Spring Thickness Of	The BEST Spring Length Is	Max Turns Using BEST Length	Max Turns If Spring Is 10" Short Or Long
0.26	0.010	110.8	23.6	23.5	0.32	0.010	109.5	21.3	21.2
0.26	0.011	100.8	21.5	21.3	0.32	0.011	99.5	19.4	19.3
0.26	0.012	92.4	19.7	19.5	0.32	0.012	91.2	17.8	17.6
0.26	0.013	85.3	18.2	18.0	0.32	0.013	84.2	16.4	16.3
0.26	0.014	79.2	16.9	16.7	0.32	0.014	78.2	15.2	15.1
0.26	0.015	73.9	15.7	15.6	0.32	0.015	73.0	14.2	14.0
0.26	0.016	69.3	14.8	14.6	0.32	0.016	68.4	13.3	13.1
0.26	0.017	65.2	13.9	13.7	0.32	0.017	64.4	12.5	12.4
0.27	0.010	110.6	23.2	23.1	0.33	0.010	109.2	20.95	20.84
0.27	0.011	100.6	21.1	21.0	0.33	0.011	99.3	19.05	18.93
0.27	0.012	92.2	19.3	19.2	0.33	0.012	91.0	17.46	17.33
0.27	0.013	85.1	17.9	17.7	0.33	0.013	84.0	16.12	15.97
0.27	0.014	79.0	16.6	16.4	0.33	0.014	78.0	14.97	14.81
0.27	0.015	73.8	15.5	15.3	0.33	0.015	72.8	13.97	13.80
0.27	0.016	69.1	14.5	14.3	0.33	0.016	68.3	13.10	12.92
0.27	0.017	65.1	13.7	13.5	0.33	0.017	64.2	12.32	12.14
0.28	0.010	110.4	22.8	22.7	0.34	0.010	109.0	20.6	20.5
0.28	0.011	100.4	20.8	20.6	0.34	0.011	99.0	18.7	18.6
0.28	0.012	92.0	19.0	18.9	0.34	0.012	90.8	17.2	17.0
0.28	0.013	84.9	17.6	17.4	0.34	0.013	83.8	15.8	15.7
0.28	0.014	78.9	16.3	16.1	0.34	0.014	77.8	14.7	14.6
0.28	0.015	73.6	15.2	15.0	0.34	0.015	72.6	13.7	13.6
0.28	0.016	69.0	14.3	14.1	0.34	0.016	68.1	12.9	12.7
0.28	0.017	64.9	13.4	13.2	0.34	0.017	64.1	12.1	11.9
0.29	0.010	110.2	22.4	22.3	0.35	0.010	108.7	20.2	20.1
0.29	0.011	100.2	20.4	20.3	0.35	0.011	98.8	18.4	18.3
0.29	0.012	91.8	18.7	18.6	0.35	0.012	90.6	16.9	16.7
0.29	0.013	84.8	17.3	17.1	0.35	0.013	83.6	15.6	15.4
0.29	0.014	78.7	16.0	15.9	0.35	0.014	77.6	14.4	14.3
0.29	0.015	73.5	15.0	14.8	0.35	0.015	72.5	13.5	13.3
0.29	0.016	68.9	14.0	13.8	0.35	0.016	67.9	12.6	12.5
0.29	0.017	64.8	13.2	13.0	0.35	0.017	63.9	11.9	11.7
0.30	0.010	110.0	22.1	22.0	0.36	0.010	108.4	19.9	19.8
0.30	0.011	100.0	20.1	19.9	0.36	0.011	98.5	18.1	17.9
0.30	0.012	91.6	18.4	18.3	0.36	0.012	90.3	16.6	16.4
0.30	0.013	84.6	17.0	16.8	0.36	0.013	83.4	15.3	15.1
0.30	0.014	78.5	15.8	15.6	0.36	0.014	77.4	14.2	14.0
0.30	0.015	73.3	14.7	14.5	0.36	0.015	72.3	13.2	13.1
0.30	0.016	68.7	13.8	13.6	0.36	0.016	67.8	12.4	12.2
0.30	0.017	64.7	13.0	12.8	0.36	0.017	63.8	11.7	11.5
0.31	0.010	109.7	21.7	21.6	0.37	0.010	108.1	19.5	19.4
0.31	0.011	99.7	19.7	19.6	0.37	0.011	98.3	17.7	17.6
0.31	0.012	91.4	18.1	17.9	0.37	0.012	90.1	16.3	16.1
0.31	0.013	84.4	16.7	16.5	0.37	0.013	83.2	15.0	14.9
0.31	0.014	78.4	15.5	15.3	0.37	0.014	77.2	13.9	13.8
0.31	0.015	73.1	14.5	14.3	0.37	0.015	72.1	13.0	12.9
0.31	0.016	68.6	13.6	13.4	0.37	0.016	67.6	12.2	12.0
0.31	0.017	64.5	12.8	12.6	0.37	0.017	63.6	11.5	11.3

Chart A-1.70 (Continued): Only For Barrels With An Inside Diameter Of 1.70 Inches

(Arbor diameters of from 0.38 Inches thru 0.49 Inches, and spring thicknesses of from 0.010 Thru 0.017 Inches)

For An Arbor Diameter Of	And A Spring Thickness Of	The BEST Spring Length Is	Max Turns Using BEST Length	Max Turns If Spring Is 10" Short Or Long	For An Arbor Diameter Of	And A Spring Thickness Of	The BEST Spring Length Is	Max Turns Using BEST Length	Max Turns If Spring Is 10" Short Or Long
0.38	0.010	107.8	19.2	19.1	0.44	0.010	105.9	17.2	17.1
0.38	0.011	98.0	17.4	17.3	0.44	0.011	96.3	15.6	15.5
0.38	0.012	89.8	16.0	15.8	0.44	0.012	88.2	14.3	14.2
0.38	0.013	82.9	14.7	14.6	0.44	0.013	81.5	13.2	13.1
0.38	0.014	77.0	13.7	13.5	0.44	0.014	75.6	12.3	12.1
0.38	0.015	71.9	12.8	12.6	0.44	0.015	70.6	11.4	11.3
0.38	0.016	67.4	12.0	11.8	0.44	0.016	66.2	10.7	10.6
0.38	0.017	63.4	11.3	11.1	0.44	0.017	62.3	10.1	9.9
0.39	0.010	107.5	18.8	18.7	0.45	0.010	105.5	16.8	16.7
0.39	0.011	97.7	17.1	17.0	0.45	0.011	95.9	15.3	15.2
0.39	0.012	89.6	15.7	15.6	0.45	0.012	87.9	14.0	13.9
0.39	0.013	82.7	14.5	14.3	0.45	0.013	81.2	13.0	12.8
0.39	0.014	76.8	13.5	13.3	0.45	0.014	75.4	12.0	11.9
0.39	0.015	71.7	12.6	12.4	0.45	0.015	70.4	11.2	11.1
0.39	0.016	67.2	11.8	11.6	0.45	0.016	66.0	10.5	10.4
0.39	0.017	63.2	11.1	10.9	0.45	0.017	62.1	9.9	9.7
0.40	0.010	107.2	18.5	18.4	0.46	0.010	105.2	16.5	16.4
0.40	0.011	97.5	16.8	16.7	0.46	0.011	95.6	15.0	14.9
0.40	0.012	89.3	15.4	15.3	0.46	0.012	87.7	13.8	13.6
0.40	0.013	82.5	14.2	14.1	0.46	0.013	80.9	12.7	12.6
0.40	0.014	76.6	13.2	13.1	0.46	0.014	75.1	11.8	11.7
0.40	0.015	71.5	12.3	12.2	0.46	0.015	70.1	11.0	10.9
0.40	0.016	67.0	11.6	11.4	0.46	0.016	65.7	10.3	10.2
0.40	0.017	63.1	10.9	10.7	0.46	0.017	61.9	9.7	9.5
0.41	0.010	106.9	18.15	18	0.47	0.010	104.8	16.2	16.1
0.41	0.011	97.2	16.50	16	0.47	0.011	95.3	14.7	14.6
0.41	0.012	89.1	15.13	15	0.47	0.012	87.3	13.5	13.4
0.41	0.013	82.2	13.97	14	0.47	0.013	80.6	12.5	12.3
0.41	0.014	76.3	12.97	13	0.47	0.014	74.9	11.6	11.4
0.41	0.015	71.3	12.10	12	0.47	0.015	69.9	10.8	10.7
0.41	0.016	66.8	11.35	11	0.47	0.016	65.5	10.1	10.0
0.41	0.017	62.9	10.68	10	0.47	0.017	61.7	9.5	9.4
0.42	0.010	106.6	17.8	17.7	0.48	0.010	104.4	15.9	15.8
0.42	0.011	96.9	16.2	16.1	0.48	0.011	94.9	14.5	14.3
0.42	0.012	88.8	14.9	14.7	0.48	0.012	87.0	13.3	13.1
0.42	0.013	82.0	13.7	13.6	0.48	0.013	80.3	12.2	12.1
0.42	0.014	76.1	12.7	12.6	0.48	0.014	74.6	11.4	11.2
0.42	0.015	71.0	11.9	11.7	0.48	0.015	69.6	10.6	10.4
0.42	0.016	66.6	11.1	11.0	0.48	0.016	65.3	9.9	9.8
0.42	0.017	62.7	10.5	10.3	0.48	0.017	61.4	9.4	9.2
0.43	0.010	106.2	17.5	17.4	0.49	0.010	104.1	15.60	15.50
0.43	0.011	96.6	15.9	15.8	0.49	0.011	94.6	14.18	14.07
0.43	0.012	88.5	14.6	14.5	0.49	0.012	86.7	13.00	12.88
0.43	0.013	81.7	13.5	13.3	0.49	0.013	80.0	12.00	11.87
0.43	0.014	75.9	12.5	12.3	0.49	0.014	74.3	11.14	11.00
0.43	0.015	70.8	11.7	11.5	0.49	0.015	69.4	10.40	10.25
0.43	0.016	66.4	10.9	10.8	0.49	0.016	65.0	9.75	9.58
0.43	0.017	62.5	10.3	10.1	0.49	0.017	61.2	9.18	9.00

Chart A-1.75: Only For Barrels With An Inside Diameter Of 1.75 Inches

(Arbor diameters of from 0.26 Inches thru 0.37 Inches, and spring thicknesses of from 0.010 Thru 0.017 Inches)

For An Arbor Diameter Of	And A Spring Thickness Of	The BEST Spring Length Is	Max Turns Using BEST Length	Max Turns If Spring Is 10" Short Or Long	For An Arbor Diameter Of	And A Spring Thickness Of	The BEST Spring Length Is	Max Turns Using BEST Length	Max Turns If Spring Is 10" Short Or Long
0.26	0.010	117.6	24.6	24.5	0.32	0.010	116.2	22.3	22.2
0.26	0.011	106.9	22.4	22.3	0.32	0.011	105.7	20.3	20.2
0.26	0.012	98.0	20.5	20.4	0.32	0.012	96.9	18.6	18.5
0.26	0.013	90.5	18.9	18.8	0.32	0.013	89.4	17.2	17.0
0.26	0.014	84.0	17.6	17.4	0.32	0.014	83.0	15.9	15.8
0.26	0.015	78.4	16.4	16.2	0.32	0.015	77.5	14.9	14.7
0.26	0.016	73.5	15.4	15.2	0.32	0.016	72.7	13.9	13.8
0.26	0.017	69.2	14.5	14.3	0.32	0.017	68.4	13.1	12.9
0.27	0.010	117.4	24.2	24.1	0.33	0.010	116.0	21.9	21.8
0.27	0.011	106.7	22.0	21.9	0.33	0.011	105.4	19.9	19.8
0.27	0.012	97.8	20.2	20.0	0.33	0.012	96.7	18.3	18.1
0.27	0.013	90.3	18.6	18.5	0.33	0.013	89.2	16.9	16.7
0.27	0.014	83.9	17.3	17.1	0.33	0.014	82.8	15.7	15.5
0.27	0.015	78.3	16.1	16.0	0.33	0.015	77.3	14.6	14.5
0.27	0.016	73.4	15.1	15.0	0.33	0.016	72.5	13.7	13.5
0.27	0.017	69.1	14.2	14.1	0.33	0.017	68.2	12.9	12.7
0.28	0.010	117.2	23.8	23.7	0.34	0.010	115.7	21.6	21.5
0.28	0.011	106.5	21.7	21.5	0.34	0.011	105.2	19.6	19.5
0.28	0.012	97.7	19.8	19.7	0.34	0.012	96.4	18.0	17.8
0.28	0.013	90.1	18.3	18.2	0.34	0.013	89.0	16.6	16.5
0.28	0.014	83.7	17.0	16.9	0.34	0.014	82.7	15.4	15.3
0.28	0.015	78.1	15.9	15.7	0.34	0.015	77.1	14.4	14.2
0.28	0.016	73.2	14.9	14.7	0.34	0.016	72.3	13.5	13.3
0.28	0.017	68.9	14.0	13.8	0.34	0.017	68.1	12.7	12.5
0.29	0.010	117.0	23.4	23.3	0.35	0.010	115.5	21.2	21.1
0.29	0.011	106.3	21.3	21.2	0.35	0.011	105.0	19.3	19.2
0.29	0.012	97.5	19.5	19.4	0.35	0.012	96.2	17.7	17.5
0.29	0.013	90.0	18.0	17.9	0.35	0.013	88.8	16.3	16.2
0.29	0.014	83.5	16.7	16.6	0.35	0.014	82.5	15.1	15.0
0.29	0.015	78.0	15.6	15.5	0.35	0.015	77.0	14.1	14.0
0.29	0.016	73.1	14.6	14.5	0.35	0.016	72.2	13.2	13.1
0.29	0.017	68.8	13.8	13.6	0.35	0.017	67.9	12.5	12.3
0.30	0.010	116.7	23.0	22.9	0.36	0.010	115.2	20.8	20.7
0.30	0.011	106.1	21.0	20.8	0.36	0.011	104.7	18.9	18.8
0.30	0.012	97.3	19.2	19.1	0.36	0.012	96.0	17.4	17.2
0.30	0.013	89.8	17.7	17.6	0.36	0.013	88.6	16.0	15.9
0.30	0.014	83.4	16.5	16.3	0.36	0.014	82.3	14.9	14.7
0.30	0.015	77.8	15.4	15.2	0.36	0.015	76.8	13.9	13.7
0.30	0.016	73.0	14.4	14.2	0.36	0.016	72.0	13.0	12.9
0.30	0.017	68.7	13.6	13.4	0.36	0.017	67.7	12.3	12.1
0.31	0.010	116.5	22.7	22.6	0.37	0.010	114.9	20.5	20.4
0.31	0.011	105.9	20.6	20.5	0.37	0.011	104.4	18.6	18.5
0.31	0.012	97.1	18.9	18.8	0.37	0.012	95.7	17.1	16.9
0.31	0.013	89.6	17.4	17.3	0.37	0.013	88.4	15.8	15.6
0.31	0.014	83.2	16.2	16.0	0.37	0.014	82.1	14.6	14.5
0.31	0.015	77.7	15.1	15.0	0.37	0.015	76.6	13.7	13.5
0.31	0.016	72.8	14.2	14.0	0.37	0.016	71.8	12.8	12.6
0.31	0.017	68.5	13.3	13.2	0.37	0.017	67.6	12.0	11.9

Chart A-1.75 (Continued): Only For Barrels With An Inside Diameter Of 1.75 Inches

(Arbor diameters of from 0.38 Inches thru 0.49 Inches, and spring thicknesses of from 0.010 Thru 0.017 Inches)

For An Arbor Diameter Of	And A Spring Thickness Of	The BEST Spring Length Is	Max Turns Using BEST Length	Max Turns If Spring Is 10" Short Or Long	For An Arbor Diameter Of	And A Spring Thickness Of	The BEST Spring Length Is	Max Turns Using BEST Length	Max Turns If Spring Is 10" Short Or Long
0.38	0.010	114.6	20.1	20.0	0.44	0.010	112.7	18.1	18.0
0.38	0.011	104.2	18.3	18.2	0.44	0.011	102.4	16.5	16.3
0.38	0.012	95.5	16.8	16.7	0.44	0.012	93.9	15.1	15.0
0.38	0.013	88.1	15.5	15.4	0.44	0.013	86.7	13.9	13.8
0.38	0.014	81.9	14.4	14.2	0.44	0.014	80.5	12.9	12.8
0.38	0.015	76.4	13.4	13.3	0.44	0.015	75.1	12.1	11.9
0.38	0.016	71.6	12.6	12.4	0.44	0.016	70.4	11.3	11.2
0.38	0.017	67.4	11.8	11.7	0.44	0.017	66.3	10.6	10.5
0.39	0.010	114.3	19.8	19.7	0.45	0.010	112.3	17.8	17.7
0.39	0.011	103.9	18.0	17.9	0.45	0.011	102.1	16.2	16.0
0.39	0.012	95.2	16.5	16.4	0.45	0.012	93.6	14.8	14.7
0.39	0.013	87.9	15.2	15.1	0.45	0.013	86.4	13.7	13.5
0.39	0.014	81.6	14.1	14.0	0.45	0.014	80.2	12.7	12.6
0.39	0.015	76.2	13.2	13.0	0.45	0.015	74.9	11.8	11.7
0.39	0.016	71.4	12.4	12.2	0.45	0.016	70.2	11.1	10.9
0.39	0.017	67.2	11.6	11.5	0.45	0.017	66.1	10.5	10.3
0.40	0.010	114.0	19.4	19.3	0.46	0.010	112.0	17.4	17.4
0.40	0.011	103.6	17.7	17.6	0.46	0.011	101.8	15.9	15.8
0.40	0.012	95.0	16.2	16.1	0.46	0.012	93.3	14.5	14.4
0.40	0.013	87.7	15.0	14.8	0.46	0.013	86.1	13.4	13.3
0.40	0.014	81.4	13.9	13.7	0.46	0.014	80.0	12.5	12.3
0.40	0.015	76.0	13.0	12.8	0.46	0.015	74.6	11.6	11.5
0.40	0.016	71.2	12.1	12.0	0.46	0.016	70.0	10.9	10.7
0.40	0.017	67.0	11.4	11.3	0.46	0.017	65.9	10.3	10.1
0.41	0.010	113.7	19.1	19.0	0.47	0.010	111.6	17.1	17.0
0.41	0.011	103.3	17.4	17.2	0.47	0.011	101.4	15.6	15.5
0.41	0.012	94.7	15.9	15.8	0.47	0.012	93.0	14.3	14.2
0.41	0.013	87.4	14.7	14.6	0.47	0.013	85.8	13.2	13.1
0.41	0.014	81.2	13.6	13.5	0.47	0.014	79.7	12.2	12.1
0.41	0.015	75.8	12.7	12.6	0.47	0.015	74.4	11.4	11.3
0.41	0.016	71.0	11.9	11.8	0.47	0.016	69.7	10.7	10.6
0.41	0.017	66.9	11.2	11.1	0.47	0.017	65.6	10.1	9.9
0.42	0.010	113.3	18.8	18.7	0.48	0.010	111.2	16.8	16.7
0.42	0.011	103.0	17.1	16.9	0.48	0.011	101.1	15.3	15.2
0.42	0.012	94.4	15.6	15.5	0.48	0.012	92.7	14.0	13.9
0.42	0.013	87.2	14.4	14.3	0.48	0.013	85.6	12.9	12.8
0.42	0.014	81.0	13.4	13.3	0.48	0.014	79.4	12.0	11.9
0.42	0.015	75.6	12.5	12.4	0.48	0.015	74.1	11.2	11.1
0.42	0.016	70.8	11.7	11.6	0.48	0.016	69.5	10.5	10.4
0.42	0.017	66.7	11.0	10.9	0.48	0.017	65.4	9.9	9.7
0.43	0.010	113.0	18.4	18.3	0.49	0.010	110.8	16.5	16.4
0.43	0.011	102.7	16.7	16.6	0.49	0.011	100.8	15.0	14.9
0.43	0.012	94.2	15.4	15.2	0.49	0.012	92.4	13.8	13.6
0.43	0.013	86.9	14.2	14.0	0.49	0.013	85.3	12.7	12.6
0.43	0.014	80.7	13.2	13.0	0.49	0.014	79.2	11.8	11.7
0.43	0.015	75.3	12.3	12.1	0.49	0.015	73.9	11.0	10.9
0.43	0.016	70.6	11.5	11.4	0.49	0.016	69.3	10.3	10.2
0.43	0.017	66.5	10.8	10.7	0.49	0.017	65.2	9.7	9.5

Chart A-1.80: Only For Barrels With An Inside Diameter Of 1.80 Inches

(Arbor diameters of from 0.26 Inches thru 0.37 Inches, and spring thicknesses of from 0.010 Thru 0.017 Inches)

For An Arbor Diameter Of	And A Spring Thickness Of	The BEST Spring Length Is	Max Turns Using BEST Length	Max Turns If Spring Is 10" Short Or Long	For An Arbor Diameter Of	And A Spring Thickness Of	The BEST Spring Length Is	Max Turns Using BEST Length	Max Turns If Spring Is 10" Short Or Long
0.26	0.010	124.6	25.6	25.5	0.32	0.010	123.2	23.3	23.2
0.26	0.011	113.3	23.3	23.2	0.32	0.011	112.0	21.2	21.1
0.26	0.012	103.8	21.3	21.2	0.32	0.012	102.7	19.4	19.3
0.26	0.013	95.8	19.7	19.6	0.32	0.013	94.8	17.9	17.8
0.26	0.014	89.0	18.3	18.2	0.32	0.014	88.0	16.6	16.5
0.26	0.015	83.1	17.1	16.9	0.32	0.015	82.1	15.5	15.4
0.26	0.016	77.9	16.0	15.8	0.32	0.016	77.0	14.5	14.4
0.26	0.017	73.3	15.1	14.9	0.32	0.017	72.5	13.7	13.5
0.27	0.010	124.4	25.2	25.1	0.33	0.010	123.0	22.9	22.8
0.27	0.011	113.1	22.9	22.8	0.33	0.011	111.8	20.8	20.7
0.27	0.012	103.6	21.0	20.9	0.33	0.012	102.5	19.1	19.0
0.27	0.013	95.7	19.4	19.3	0.33	0.013	94.6	17.6	17.5
0.27	0.014	88.8	18.0	17.9	0.33	0.014	87.8	16.4	16.2
0.27	0.015	82.9	16.8	16.7	0.33	0.015	82.0	15.3	15.1
0.27	0.016	77.7	15.8	15.6	0.33	0.016	76.8	14.3	14.2
0.27	0.017	73.2	14.8	14.7	0.33	0.017	72.3	13.5	13.3
0.28	0.010	124.2	24.8	24.7	0.34	0.010	122.7	22.5	22.4
0.28	0.011	112.9	22.6	22.4	0.34	0.011	111.5	20.5	20.4
0.28	0.012	103.5	20.7	20.6	0.34	0.012	102.2	18.8	18.7
0.28	0.013	95.5	19.1	19.0	0.34	0.013	94.4	17.3	17.2
0.28	0.014	88.7	17.7	17.6	0.34	0.014	87.6	16.1	16.0
0.28	0.015	82.8	16.5	16.4	0.34	0.015	81.8	15.0	14.9
0.28	0.016	77.6	15.5	15.4	0.34	0.016	76.7	14.1	13.9
0.28	0.017	73.0	14.6	14.4	0.34	0.017	72.2	13.3	13.1
0.29	0.010	123.9	24.4	24.3	0.35	0.010	122.4	22.2	22.1
0.29	0.011	112.7	22.2	22.1	0.35	0.011	111.3	20.1	20.0
0.29	0.012	103.3	20.4	20.2	0.35	0.012	102.0	18.5	18.4
0.29	0.013	95.3	18.8	18.7	0.35	0.013	94.2	17.0	16.9
0.29	0.014	88.5	17.4	17.3	0.35	0.014	87.4	15.8	15.7
0.29	0.015	82.6	16.3	16.1	0.35	0.015	81.6	14.8	14.6
0.29	0.016	77.5	15.3	15.1	0.35	0.016	76.5	13.9	13.7
0.29	0.017	72.9	14.4	14.2	0.35	0.017	72.0	13.0	12.9
0.30	0.010	123.7	24.0	23.9	0.36	0.010	122.1	21.8	21.7
0.30	0.011	112.5	21.8	21.7	0.36	0.011	111.0	19.8	19.7
0.30	0.012	103.1	20.0	19.9	0.36	0.012	101.8	18.2	18.1
0.30	0.013	95.2	18.5	18.4	0.36	0.013	94.0	16.8	16.6
0.30	0.014	88.4	17.2	17.0	0.36	0.014	87.2	15.6	15.4
0.30	0.015	82.5	16.0	15.9	0.36	0.015	81.4	14.5	14.4
0.30	0.016	77.3	15.0	14.9	0.36	0.016	76.3	13.6	13.5
0.30	0.017	72.8	14.1	14.0	0.36	0.017	71.9	12.8	12.7
0.31	0.010	123.5	23.7	23.6	0.37	0.010	121.9	21.4	21.3
0.31	0.011	112.2	21.5	21.4	0.37	0.011	110.8	19.5	19.4
0.31	0.012	102.9	19.7	19.6	0.37	0.012	101.5	17.9	17.8
0.31	0.013	95.0	18.2	18.1	0.37	0.013	93.7	16.5	16.4
0.31	0.014	88.2	16.9	16.8	0.37	0.014	87.0	15.3	15.2
0.31	0.015	82.3	15.8	15.6	0.37	0.015	81.2	14.3	14.2
0.31	0.016	77.2	14.8	14.6	0.37	0.016	76.2	13.4	13.3
0.31	0.017	72.6	13.9	13.8	0.37	0.017	71.7	12.6	12.5

Chart A-1.80 (Continued): Only For Barrels With An Inside Diameter Of 1.80 Inches

(Arbor diameters of from 0.38 Inches thru 0.49 Inches, and spring thicknesses of from 0.010 Thru 0.017 Inches)

For An Arbor Diameter Of	And A Spring Thickness Of	The BEST Spring Length Is	Max Turns Using BEST Length	Max Turns If Spring Is 10" Short Or Long	For An Arbor Diameter Of	And A Spring Thickness Of	The BEST Spring Length Is	Max Turns Using BEST Length	Max Turns If Spring Is 10" Short Or Long
0.38	0.010	121.6	21.1	21.0	0.44	0.010	119.6	19.0	18.9
0.38	0.011	110.5	19.2	19.1	0.44	0.011	108.8	17.3	17.2
0.38	0.012	101.3	17.6	17.5	0.44	0.012	99.7	15.9	15.7
0.38	0.013	93.5	16.2	16.1	0.44	0.013	92.0	14.6	14.5
0.38	0.014	86.8	15.1	14.9	0.44	0.014	85.5	13.6	13.5
0.38	0.015	81.0	14.1	13.9	0.44	0.015	79.8	12.7	12.5
0.38	0.016	76.0	13.2	13.0	0.44	0.016	74.8	11.9	11.7
0.38	0.017	71.5	12.4	12.2	0.44	0.017	70.4	11.2	11.0
0.39	0.010	121.3	20.7	20.6	0.45	0.010	119.3	18.7	18.6
0.39	0.011	110.2	18.8	18.7	0.45	0.011	108.4	17.0	16.9
0.39	0.012	101.1	17.3	17.2	0.45	0.012	99.4	15.6	15.5
0.39	0.013	93.3	15.9	15.8	0.45	0.013	91.8	14.4	14.3
0.39	0.014	86.6	14.8	14.7	0.45	0.014	85.2	13.4	13.2
0.39	0.015	80.8	13.8	13.7	0.45	0.015	79.5	12.5	12.3
0.39	0.016	75.8	13.0	12.8	0.45	0.016	74.6	11.7	11.5
0.39	0.017	71.3	12.2	12.0	0.45	0.017	70.2	11.0	10.8
0.40	0.010	121.0	20.4	20.3	0.46	0.010	118.9	18.4	18.3
0.40	0.011	110.0	18.5	18.4	0.46	0.011	108.1	16.7	16.6
0.40	0.012	100.8	17.0	16.9	0.46	0.012	99.1	15.3	15.2
0.40	0.013	93.0	15.7	15.6	0.46	0.013	91.5	14.1	14.0
0.40	0.014	86.4	14.6	14.4	0.46	0.014	84.9	13.1	13.0
0.40	0.015	80.6	13.6	13.5	0.46	0.015	79.3	12.2	12.1
0.40	0.016	75.6	12.7	12.6	0.46	0.016	74.3	11.5	11.3
0.40	0.017	71.1	12.0	11.8	0.46	0.017	70.0	10.8	10.7
0.41	0.010	120.6	20.0	19.9	0.47	0.010	118.6	18.0	18.0
0.41	0.011	109.7	18.2	18.1	0.47	0.011	107.8	16.4	16.3
0.41	0.012	100.5	16.7	16.6	0.47	0.012	98.8	15.0	14.9
0.41	0.013	92.8	15.4	15.3	0.47	0.013	91.2	13.9	13.8
0.41	0.014	86.2	14.3	14.2	0.47	0.014	84.7	12.9	12.8
0.41	0.015	80.4	13.4	13.2	0.47	0.015	79.0	12.0	11.9
0.41	0.016	75.4	12.5	12.4	0.47	0.016	74.1	11.3	11.1
0.41	0.017	71.0	11.8	11.6	0.47	0.017	69.7	10.6	10.5
0.42	0.010	120.3	19.7	19.6	0.48	0.010	118.2	17.7	17.6
0.42	0.011	109.4	17.9	17.8	0.48	0.011	107.4	16.1	16.0
0.42	0.012	100.3	16.4	16.3	0.48	0.012	98.5	14.8	14.7
0.42	0.013	92.5	15.2	15.0	0.48	0.013	90.9	13.6	13.5
0.42	0.014	85.9	14.1	13.9	0.48	0.014	84.4	12.7	12.5
0.42	0.015	80.2	13.1	13.0	0.48	0.015	78.8	11.8	11.7
0.42	0.016	75.2	12.3	12.2	0.48	0.016	73.9	11.1	10.9
0.42	0.017	70.8	11.6	11.4	0.48	0.017	69.5	10.4	10.3
0.43	0.010	120.0	19.4	19.3	0.49	0.010	117.8	17.4	17.3
0.43	0.011	109.1	17.6	17.5	0.49	0.011	107.1	15.8	15.7
0.43	0.012	100.0	16.1	16.0	0.49	0.012	98.2	14.5	14.4
0.43	0.013	92.3	14.9	14.8	0.49	0.013	90.6	13.4	13.3
0.43	0.014	85.7	13.8	13.7	0.49	0.014	84.1	12.4	12.3
0.43	0.015	80.0	12.9	12.8	0.49	0.015	78.5	11.6	11.5
0.43	0.016	75.0	12.1	12.0	0.49	0.016	73.6	10.9	10.7
0.43	0.017	70.6	11.4	11.2	0.49	0.017	69.3	10.2	10.1

Selecting the Perfect Spring for Your Barreled Spring Clock

Chart A-1.85: Only For Barrels With An Inside Diameter Of 1.85 Inches

(Arbor diameters of from 0.26 Inches thru 0.37 Inches, and spring thicknesses of from 0.010 Thru 0.017 Inches)

For An Arbor Diameter Of	And A Spring Thickness Of	The BEST Spring Length Is	Max Turns Using BEST Length	Max Turns If Spring Is 10" Short Or Long	For An Arbor Diameter Of	And A Spring Thickness Of	The BEST Spring Length Is	Max Turns Using BEST Length	Max Turns If Spring Is 10" Short Or Long
0.26	0.010	131.7	26.6	26.5	0.32	0.010	130.4	24.3	24.2
0.26	0.011	119.8	24.2	24.1	0.32	0.011	118.5	22.1	22.0
0.26	0.012	109.8	22.2	22.1	0.32	0.012	108.6	20.2	20.1
0.26	0.013	101.3	20.5	20.3	0.32	0.013	100.3	18.7	18.5
0.26	0.014	94.1	19.0	18.9	0.32	0.014	93.1	17.3	17.2
0.26	0.015	87.8	17.7	17.6	0.32	0.015	86.9	16.2	16.0
0.26	0.016	82.3	16.6	16.5	0.32	0.016	81.5	15.2	15.0
0.26	0.017	77.5	15.6	15.5	0.32	0.017	76.7	14.3	14.1
0.27	0.010	131.5	26.2	26.1	0.33	0.010	130.1	23.9	23.8
0.27	0.011	119.6	23.8	23.7	0.33	0.011	118.3	21.7	21.6
0.27	0.012	109.6	21.8	21.7	0.33	0.012	108.4	19.9	19.8
0.27	0.013	101.2	20.2	20.0	0.33	0.013	100.1	18.4	18.3
0.27	0.014	94.0	18.7	18.6	0.33	0.014	92.9	17.1	16.9
0.27	0.015	87.7	17.5	17.3	0.33	0.015	86.7	15.9	15.8
0.27	0.016	82.2	16.4	16.2	0.33	0.016	81.3	14.9	14.8
0.27	0.017	77.4	15.4	15.3	0.33	0.017	76.5	14.0	13.9
0.28	0.010	131.3	25.8	25.7	0.34	0.010	129.9	23.5	23.4
0.28	0.011	119.4	23.5	23.4	0.34	0.011	118.1	21.4	21.3
0.28	0.012	109.4	21.5	21.4	0.34	0.012	108.2	19.6	19.5
0.28	0.013	101.0	19.8	19.7	0.34	0.013	99.9	18.1	18.0
0.28	0.014	93.8	18.4	18.3	0.34	0.014	92.8	16.8	16.7
0.28	0.015	87.5	17.2	17.1	0.34	0.015	86.6	15.7	15.5
0.28	0.016	82.1	16.1	16.0	0.34	0.016	81.2	14.7	14.6
0.28	0.017	77.2	15.2	15.0	0.34	0.017	76.4	13.8	13.7
0.29	0.010	131.1	25.4	25.3	0.35	0.010	129.6	23.1	23.0
0.29	0.011	119.2	23.1	23.0	0.35	0.011	117.8	21.0	20.9
0.29	0.012	109.2	21.2	21.1	0.35	0.012	108.0	19.3	19.2
0.29	0.013	100.8	19.5	19.4	0.35	0.013	99.7	17.8	17.7
0.29	0.014	93.6	18.2	18.0	0.35	0.014	92.6	16.5	16.4
0.29	0.015	87.4	16.9	16.8	0.35	0.015	86.4	15.4	15.3
0.29	0.016	81.9	15.9	15.7	0.35	0.016	81.0	14.5	14.3
0.29	0.017	77.1	14.9	14.8	0.35	0.017	76.2	13.6	13.5
0.30	0.010	130.9	25.0	24.9	0.36	0.010	129.3	22.8	22.7
0.30	0.011	119.0	22.7	22.7	0.36	0.011	117.6	20.7	20.6
0.30	0.012	109.1	20.9	20.7	0.36	0.012	107.8	19.0	18.9
0.30	0.013	100.7	19.2	19.1	0.36	0.013	99.5	17.5	17.4
0.30	0.014	93.5	17.9	17.8	0.36	0.014	92.4	16.3	16.1
0.30	0.015	87.2	16.7	16.6	0.36	0.015	86.2	15.2	15.1
0.30	0.016	81.8	15.6	15.5	0.36	0.016	80.8	14.2	14.1
0.30	0.017	77.0	14.7	14.6	0.36	0.017	76.1	13.4	13.2
0.31	0.010	130.6	24.6	24.6	0.37	0.010	129.0	22.4	22.3
0.31	0.011	118.8	22.4	22.3	0.37	0.011	117.3	20.4	20.3
0.31	0.012	108.9	20.5	20.4	0.37	0.012	107.5	18.7	18.6
0.31	0.013	100.5	19.0	18.8	0.37	0.013	99.3	17.2	17.1
0.31	0.014	93.3	17.6	17.5	0.37	0.014	92.2	16.0	15.9
0.31	0.015	87.1	16.4	16.3	0.37	0.015	86.0	14.9	14.8
0.31	0.016	81.6	15.4	15.3	0.37	0.016	80.6	14.0	13.9
0.31	0.017	76.8	14.5	14.3	0.37	0.017	75.9	13.2	13.0

Chart A-1.85 (Continued): Only For Barrels With An Inside Diameter Of 1.85 Inches

(Arbor diameters of from 0.38 Inches thru 0.49 Inches, and spring thicknesses of from 0.010 Thru 0.017 Inches)

For An Arbor Diameter Of	And A Spring Thickness Of	The BEST Spring Length Is	Max Turns Using BEST Length	Max Turns If Spring Is 10" Short Or Long	For An Arbor Diameter Of	And A Spring Thickness Of	The BEST Spring Length Is	Max Turns Using BEST Length	Max Turns If Spring Is 10" Short Or Long
0.38	0.010	128.7	22.0	22.0	0.44	0.010	126.8	20.0	19.9
0.38	0.011	117.0	20.0	19.9	0.44	0.011	115.3	18.1	18.1
0.38	0.012	107.3	18.4	18.3	0.44	0.012	105.7	16.6	16.5
0.38	0.013	99.0	17.0	16.8	0.44	0.013	97.5	15.4	15.2
0.38	0.014	92.0	15.7	15.6	0.44	0.014	90.6	14.3	14.1
0.38	0.015	85.8	14.7	14.6	0.44	0.015	84.5	13.3	13.2
0.38	0.016	80.5	13.8	13.6	0.44	0.016	79.2	12.5	12.3
0.38	0.017	75.7	13.0	12.8	0.44	0.017	74.6	11.7	11.6
0.39	0.010	128.4	21.7	21.6	0.45	0.010	126.4	19.6	19.5
0.39	0.011	116.8	19.7	19.6	0.45	0.011	115.0	17.8	17.8
0.39	0.012	107.0	18.1	18.0	0.45	0.012	105.4	16.4	16.3
0.39	0.013	98.8	16.7	16.6	0.45	0.013	97.3	15.1	15.0
0.39	0.014	91.7	15.5	15.4	0.45	0.014	90.3	14.0	13.9
0.39	0.015	85.6	14.5	14.3	0.45	0.015	84.3	13.1	13.0
0.39	0.016	80.3	13.6	13.4	0.45	0.016	79.0	12.3	12.1
0.39	0.017	75.5	12.8	12.6	0.45	0.017	74.4	11.5	11.4
0.40	0.010	128.1	21.3	21.3	0.46	0.010	126.1	19.3	19.2
0.40	0.011	116.5	19.4	19.3	0.46	0.011	114.6	17.5	17.5
0.40	0.012	106.8	17.8	17.7	0.46	0.012	105.1	16.1	16.0
0.40	0.013	98.6	16.4	16.3	0.46	0.013	97.0	14.8	14.7
0.40	0.014	91.5	15.2	15.1	0.46	0.014	90.1	13.8	13.7
0.40	0.015	85.4	14.2	14.1	0.46	0.015	84.1	12.9	12.7
0.40	0.016	80.1	13.3	13.2	0.46	0.016	78.8	12.1	11.9
0.40	0.017	75.4	12.6	12.4	0.46	0.017	74.2	11.4	11.2
0.41	0.010	127.8	21.0	20.9	0.47	0.010	125.7	19.0	18.9
0.41	0.011	116.2	19.1	19.0	0.47	0.011	114.3	17.2	17.2
0.41	0.012	106.5	17.5	17.4	0.47	0.012	104.8	15.8	15.7
0.41	0.013	98.3	16.1	16.0	0.47	0.013	96.7	14.6	14.5
0.41	0.014	91.3	15.0	14.9	0.47	0.014	89.8	13.6	13.4
0.41	0.015	85.2	14.0	13.9	0.47	0.015	83.8	12.6	12.5
0.41	0.016	79.9	13.1	13.0	0.47	0.016	78.6	11.9	11.7
0.41	0.017	75.2	12.3	12.2	0.47	0.017	74.0	11.2	11.0
0.42	0.010	127.5	20.6	20.6	0.48	0.010	125.4	18.6	18.6
0.42	0.011	115.9	18.8	18.7	0.48	0.011	114.0	17.0	16.9
0.42	0.012	106.2	17.2	17.1	0.48	0.012	104.5	15.5	15.4
0.42	0.013	98.1	15.9	15.8	0.48	0.013	96.4	14.3	14.2
0.42	0.014	91.1	14.7	14.6	0.48	0.014	89.5	13.3	13.2
0.42	0.015	85.0	13.8	13.6	0.48	0.015	83.6	12.4	12.3
0.42	0.016	79.7	12.9	12.8	0.48	0.016	78.3	11.7	11.5
0.42	0.017	75.0	12.1	12.0	0.48	0.017	73.7	11.0	10.8
0.43	0.010	127.1	20.3	20.2	0.49	0.010	125.0	18.3	18.2
0.43	0.011	115.6	18.5	18.4	0.49	0.011	113.6	16.7	16.6
0.43	0.012	106.0	16.9	16.8	0.49	0.012	104.1	15.3	15.2
0.43	0.013	97.8	15.6	15.5	0.49	0.013	96.1	14.1	14.0
0.43	0.014	90.8	14.5	14.4	0.49	0.014	89.3	13.1	13.0
0.43	0.015	84.8	13.5	13.4	0.49	0.015	83.3	12.2	12.1
0.43	0.016	79.5	12.7	12.6	0.49	0.016	78.1	11.5	11.3
0.43	0.017	74.8	11.9	11.8	0.49	0.017	73.5	10.8	10.6

Chart A-1.90: Only For Barrels With An Inside Diameter Of 1.90 Inches

(Arbor diameters of from 0.26 Inches thru 0.37 Inches, and spring thicknesses of from 0.010 Thru 0.017 Inches)

For An Arbor Diameter Of	And A Spring Thickness Of	The BEST Spring Length Is	Max Turns Using BEST Length	Max Turns If Spring Is 10" Short Or Long	For An Arbor Diameter Of	And A Spring Thickness Of	The BEST Spring Length Is	Max Turns Using BEST Length	Max Turns If Spring Is 10" Short Or Long
0.26	0.010	139.1	27.6	27.5	0.32	0.010	137.7	25.2	25.2
0.26	0.011	126.5	25.1	25.0	0.32	0.011	125.2	22.9	22.9
0.26	0.012	115.9	23.0	22.9	0.32	0.012	114.8	21.0	20.9
0.26	0.013	107.0	21.2	21.1	0.32	0.013	106.0	19.4	19.3
0.26	0.014	99.4	19.7	19.6	0.32	0.014	98.4	18.0	17.9
0.26	0.015	92.7	18.4	18.3	0.32	0.015	91.8	16.8	16.7
0.26	0.016	86.9	17.3	17.1	0.32	0.016	86.1	15.8	15.6
0.26	0.017	81.8	16.2	16.1	0.32	0.017	81.0	14.8	14.7
0.27	0.010	138.9	27.2	27.1	0.33	0.010	137.5	24.9	24.8
0.27	0.011	126.3	24.7	24.6	0.33	0.011	125.0	22.6	22.5
0.27	0.012	115.8	22.7	22.6	0.33	0.012	114.6	20.7	20.6
0.27	0.013	106.8	20.9	20.8	0.33	0.013	105.8	19.1	19.0
0.27	0.014	99.2	19.4	19.3	0.33	0.014	98.2	17.8	17.6
0.27	0.015	92.6	18.1	18.0	0.33	0.015	91.7	16.6	16.5
0.27	0.016	86.8	17.0	16.9	0.33	0.016	85.9	15.5	15.4
0.27	0.017	81.7	16.0	15.9	0.33	0.017	80.9	14.6	14.5
0.28	0.010	138.7	26.8	26.7	0.34	0.010	137.2	24.5	24.4
0.28	0.011	126.1	24.4	24.3	0.34	0.011	124.7	22.3	22.2
0.28	0.012	115.6	22.3	22.2	0.34	0.012	114.4	20.4	20.3
0.28	0.013	106.7	20.6	20.5	0.34	0.013	105.6	18.8	18.7
0.28	0.014	99.1	19.1	19.0	0.34	0.014	98.0	17.5	17.4
0.28	0.015	92.5	17.9	17.7	0.34	0.015	91.5	16.3	16.2
0.28	0.016	86.7	16.8	16.6	0.34	0.016	85.8	15.3	15.2
0.28	0.017	81.6	15.8	15.6	0.34	0.017	80.7	14.4	14.3
0.29	0.010	138.5	26.4	26.3	0.35	0.010	137.0	24.1	24.0
0.29	0.011	125.9	24.0	23.9	0.35	0.011	124.5	21.9	21.8
0.29	0.012	115.4	22.0	21.9	0.35	0.012	114.1	20.1	20.0
0.29	0.013	106.5	20.3	20.2	0.35	0.013	105.3	18.5	18.4
0.29	0.014	98.9	18.9	18.7	0.35	0.014	97.8	17.2	17.1
0.29	0.015	92.3	17.6	17.5	0.35	0.015	91.3	16.1	16.0
0.29	0.016	86.5	16.5	16.4	0.35	0.016	85.6	15.1	14.9
0.29	0.017	81.4	15.5	15.4	0.35	0.017	80.6	14.2	14.0
0.30	0.010	138.2	26.0	25.9	0.36	0.010	136.7	23.7	23.7
0.30	0.011	125.7	23.6	23.6	0.36	0.011	124.2	21.6	21.5
0.30	0.012	115.2	21.7	21.6	0.36	0.012	113.9	19.8	19.7
0.30	0.013	106.3	20.0	19.9	0.36	0.013	105.1	18.3	18.2
0.30	0.014	98.7	18.6	18.5	0.36	0.014	97.6	17.0	16.8
0.30	0.015	92.2	17.3	17.2	0.36	0.015	91.1	15.8	15.7
0.30	0.016	86.4	16.3	16.1	0.36	0.016	85.4	14.8	14.7
0.30	0.017	81.3	15.3	15.2	0.36	0.017	80.4	14.0	13.8
0.31	0.010	138.0	25.6	25.5	0.37	0.010	136.4	23.4	23.3
0.31	0.011	125.4	23.3	23.2	0.37	0.011	124.0	21.2	21.2
0.31	0.012	115.0	21.4	21.3	0.37	0.012	113.7	19.5	19.4
0.31	0.013	106.1	19.7	19.6	0.37	0.013	104.9	18.0	17.9
0.31	0.014	98.6	18.3	18.2	0.37	0.014	97.4	16.7	16.6
0.31	0.015	92.0	17.1	17.0	0.37	0.015	90.9	15.6	15.5
0.31	0.016	86.2	16.0	15.9	0.37	0.016	85.2	14.6	14.5
0.31	0.017	81.2	15.1	14.9	0.37	0.017	80.2	13.7	13.6

Chart A-1.90 (Continued): Only For Barrels With An Inside Diameter Of 1.90 Inches

(Arbor diameters of from 0.38 Inches thru 0.49 Inches, and spring thicknesses of from 0.010 Thru 0.017 Inches)

For An Arbor Diameter Of	And A Spring Thickness Of	The BEST Spring Length Is	Max Turns Using BEST Length	Max Turns If Spring Is 10" Short Or Long	For An Arbor Diameter Of	And A Spring Thickness Of	The BEST Spring Length Is	Max Turns Using BEST Length	Max Turns If Spring Is 10" Short Or Long
0.38	0.010	136.1	23.0	22.9	0.44	0.010	134.2	20.9	20.8
0.38	0.011	123.7	20.9	20.8	0.44	0.011	122.0	19.0	18.9
0.38	0.012	113.4	19.2	19.1	0.44	0.012	111.8	17.4	17.3
0.38	0.013	104.7	17.7	17.6	0.44	0.013	103.2	16.1	16.0
0.38	0.014	97.2	16.4	16.3	0.44	0.014	95.8	14.9	14.8
0.38	0.015	90.7	15.3	15.2	0.44	0.015	89.4	13.9	13.8
0.38	0.016	85.1	14.4	14.3	0.44	0.016	83.9	13.1	12.9
0.38	0.017	80.1	13.5	13.4	0.44	0.017	78.9	12.3	12.2
0.39	0.010	135.8	22.7	22.6	0.45	0.010	133.8	20.6	20.5
0.39	0.011	123.4	20.6	20.5	0.45	0.011	121.6	18.7	18.6
0.39	0.012	113.2	18.9	18.8	0.45	0.012	111.5	17.1	17.0
0.39	0.013	104.5	17.4	17.3	0.45	0.013	102.9	15.8	15.7
0.39	0.014	97.0	16.2	16.1	0.45	0.014	95.6	14.7	14.6
0.39	0.015	90.5	15.1	15.0	0.45	0.015	89.2	13.7	13.6
0.39	0.016	84.9	14.2	14.0	0.45	0.016	83.6	12.9	12.7
0.39	0.017	79.9	13.3	13.2	0.45	0.017	78.7	12.1	12.0
0.40	0.010	135.5	22.3	22.2	0.46	0.010	133.5	20.2	20.2
0.40	0.011	123.2	20.3	20.2	0.46	0.011	121.3	18.4	18.3
0.40	0.012	112.9	18.6	18.5	0.46	0.012	111.2	16.9	16.8
0.40	0.013	104.2	17.2	17.0	0.46	0.013	102.7	15.6	15.5
0.40	0.014	96.8	15.9	15.8	0.46	0.014	95.3	14.5	14.3
0.40	0.015	90.3	14.9	14.7	0.46	0.015	89.0	13.5	13.4
0.40	0.016	84.7	13.9	13.8	0.46	0.016	83.4	12.6	12.5
0.40	0.017	79.7	13.1	13.0	0.46	0.017	78.5	11.9	11.8
0.41	0.010	135.2	21.9	21.9	0.47	0.010	133.1	19.9	19.8
0.41	0.011	122.9	19.9	19.9	0.47	0.011	121.0	18.1	18.0
0.41	0.012	112.6	18.3	18.2	0.47	0.012	110.9	16.6	16.5
0.41	0.013	104.0	16.9	16.8	0.47	0.013	102.4	15.3	15.2
0.41	0.014	96.5	15.7	15.6	0.47	0.014	95.1	14.2	14.1
0.41	0.015	90.1	14.6	14.5	0.47	0.015	88.7	13.3	13.2
0.41	0.016	84.5	13.7	13.6	0.47	0.016	83.2	12.4	12.3
0.41	0.017	79.5	12.9	12.8	0.47	0.017	78.3	11.7	11.6
0.42	0.010	134.8	21.6	21.5	0.48	0.010	132.7	19.6	19.5
0.42	0.011	122.6	19.6	19.5	0.48	0.011	120.7	17.8	17.7
0.42	0.012	112.4	18.0	17.9	0.48	0.012	110.6	16.3	16.2
0.42	0.013	103.7	16.6	16.5	0.48	0.013	102.1	15.1	15.0
0.42	0.014	96.3	15.4	15.3	0.48	0.014	94.8	14.0	13.9
0.42	0.015	89.9	14.4	14.3	0.48	0.015	88.5	13.0	12.9
0.42	0.016	84.3	13.5	13.4	0.48	0.016	82.9	12.2	12.1
0.42	0.017	79.3	12.7	12.6	0.48	0.017	78.1	11.5	11.4
0.43	0.010	134.5	21.2	21.2	0.49	0.010	132.3	19.2	19.2
0.43	0.011	122.3	19.3	19.2	0.49	0.011	120.3	17.5	17.4
0.43	0.012	112.1	17.7	17.6	0.49	0.012	110.3	16.0	15.9
0.43	0.013	103.5	16.3	16.2	0.49	0.013	101.8	14.8	14.7
0.43	0.014	96.1	15.2	15.1	0.49	0.014	94.5	13.7	13.6
0.43	0.015	89.7	14.2	14.0	0.49	0.015	88.2	12.8	12.7
0.43	0.016	84.1	13.3	13.2	0.49	0.016	82.7	12.0	11.9
0.43	0.017	79.1	12.5	12.4	0.49	0.017	77.8	11.3	11.2

Selecting the Perfect Spring for Your Barreled Spring Clock

Chart A-1.95: Only For Barrels With An Inside Diameter Of 1.95 Inches

(Arbor diameters of from 0.26 Inches thru 0.37 Inches, and spring thicknesses of from 0.010 Thru 0.017 Inches)

For An Arbor Diameter Of	And A Spring Thickness Of	The BEST Spring Length Is	Max Turns Using BEST Length	Max Turns If Spring Is 10" Short Or Long	For An Arbor Diameter Of	And A Spring Thickness Of	The BEST Spring Length Is	Max Turns Using BEST Length	Max Turns If Spring Is 10" Short Or Long
0.26	0.010	146.7	28.6	28.5	0.32	0.010	145.3	26.2	26.2
0.26	0.011	133.3	26.0	25.9	0.32	0.011	132.1	23.8	23.8
0.26	0.012	122.2	23.8	23.7	0.32	0.012	121.1	21.9	21.8
0.26	0.013	112.8	22.0	21.9	0.32	0.013	111.8	20.2	20.1
0.26	0.014	104.8	20.4	20.3	0.32	0.014	103.8	18.7	18.6
0.26	0.015	97.8	19.1	19.0	0.32	0.015	96.9	17.5	17.4
0.26	0.016	91.7	17.9	17.8	0.32	0.016	90.8	16.4	16.3
0.26	0.017	86.3	16.8	16.7	0.32	0.017	85.5	15.4	15.3
0.27	0.010	146.5	28.2	28.1	0.33	0.010	145.0	25.8	25.8
0.27	0.011	133.1	25.6	25.6	0.33	0.011	131.9	23.5	23.4
0.27	0.012	122.1	23.5	23.4	0.33	0.012	120.9	21.5	21.4
0.27	0.013	112.7	21.7	21.6	0.33	0.013	111.6	19.9	19.8
0.27	0.014	104.6	20.1	20.0	0.33	0.014	103.6	18.5	18.4
0.27	0.015	97.6	18.8	18.7	0.33	0.015	96.7	17.2	17.1
0.27	0.016	91.5	17.6	17.5	0.33	0.016	90.7	16.2	16.0
0.27	0.017	86.2	16.6	16.5	0.33	0.017	85.3	15.2	15.1
0.28	0.010	146.2	27.8	27.7	0.34	0.010	144.8	25.5	25.4
0.28	0.011	132.9	25.3	25.2	0.34	0.011	131.6	23.2	23.1
0.28	0.012	121.9	23.2	23.1	0.34	0.012	120.7	21.2	21.1
0.28	0.013	112.5	21.4	21.3	0.34	0.013	111.4	19.6	19.5
0.28	0.014	104.5	19.9	19.8	0.34	0.014	103.4	18.2	18.1
0.28	0.015	97.5	18.5	18.4	0.34	0.015	96.5	17.0	16.9
0.28	0.016	91.4	17.4	17.3	0.34	0.016	90.5	15.9	15.8
0.28	0.017	86.0	16.4	16.2	0.34	0.017	85.2	15.0	14.9
0.29	0.010	146.0	27.4	27.3	0.35	0.010	144.5	25.1	25.0
0.29	0.011	132.7	24.9	24.8	0.35	0.011	131.4	22.8	22.7
0.29	0.012	121.7	22.8	22.7	0.35	0.012	120.4	20.9	20.8
0.29	0.013	112.3	21.1	21.0	0.35	0.013	111.2	19.3	19.2
0.29	0.014	104.3	19.6	19.5	0.35	0.014	103.2	17.9	17.8
0.29	0.015	97.3	18.3	18.2	0.35	0.015	96.3	16.7	16.6
0.29	0.016	91.3	17.1	17.0	0.35	0.016	90.3	15.7	15.6
0.29	0.017	85.9	16.1	16.0	0.35	0.017	85.0	14.8	14.6
0.30	0.010	145.8	27.0	26.9	0.36	0.010	144.2	24.7	24.6
0.30	0.011	132.5	24.6	24.5	0.36	0.011	131.1	22.5	22.4
0.30	0.012	121.5	22.5	22.4	0.36	0.012	120.2	20.6	20.5
0.30	0.013	112.1	20.8	20.7	0.36	0.013	110.9	19.0	18.9
0.30	0.014	104.1	19.3	19.2	0.36	0.014	103.0	17.7	17.6
0.30	0.015	97.2	18.0	17.9	0.36	0.015	96.2	16.5	16.4
0.30	0.016	91.1	16.9	16.8	0.36	0.016	90.1	15.4	15.3
0.30	0.017	85.8	15.9	15.8	0.36	0.017	84.8	14.5	14.4
0.31	0.010	145.5	26.6	26.5	0.37	0.010	143.9	24.3	24.3
0.31	0.011	132.3	24.2	24.1	0.37	0.011	130.9	22.1	22.1
0.31	0.012	121.3	22.2	22.1	0.37	0.012	120.0	20.3	20.2
0.31	0.013	112.0	20.5	20.4	0.37	0.013	110.7	18.7	18.6
0.31	0.014	104.0	19.0	18.9	0.37	0.014	102.8	17.4	17.3
0.31	0.015	97.0	17.7	17.6	0.37	0.015	96.0	16.2	16.1
0.31	0.016	91.0	16.6	16.5	0.37	0.016	90.0	15.2	15.1
0.31	0.017	85.6	15.7	15.5	0.37	0.017	84.7	14.3	14.2

54

Chart A-1.95 (Continued): Only For Barrels With An Inside Diameter Of 1.95 Inches

(Arbor diameters of from 0.38 Inches thru 0.49 Inches, and spring thicknesses of from 0.010 Thru 0.017 Inches)

For An Arbor Diameter Of	And A Spring Thickness Of	The BEST Spring Length Is	Max Turns Using BEST Length	Max Turns If Spring Is 10" Short Or Long	For An Arbor Diameter Of	And A Spring Thickness Of	The BEST Spring Length Is	Max Turns Using BEST Length	Max Turns If Spring Is 10" Short Or Long
0.38	0.010	143.7	24.0	23.9	0.44	0.010	141.7	21.9	21.8
0.38	0.011	130.6	21.8	21.7	0.44	0.011	128.8	19.9	19.8
0.38	0.012	119.7	20.0	19.9	0.44	0.012	118.1	18.2	18.1
0.38	0.013	110.5	18.4	18.4	0.44	0.013	109.0	16.8	16.7
0.38	0.014	102.6	17.1	17.0	0.44	0.014	101.2	15.6	15.5
0.38	0.015	95.8	16.0	15.9	0.44	0.015	94.5	14.6	14.5
0.38	0.016	89.8	15.0	14.9	0.44	0.016	88.6	13.7	13.5
0.38	0.017	84.5	14.1	14.0	0.44	0.017	83.4	12.9	12.7
0.39	0.010	143.4	23.6	23.5	0.45	0.010	141.4	21.5	21.4
0.39	0.011	130.3	21.5	21.4	0.45	0.011	128.5	19.6	19.5
0.39	0.012	119.5	19.7	19.6	0.45	0.012	117.8	17.9	17.8
0.39	0.013	110.3	18.2	18.1	0.45	0.013	108.7	16.5	16.5
0.39	0.014	102.4	16.9	16.8	0.45	0.014	101.0	15.4	15.3
0.39	0.015	95.6	15.7	15.6	0.45	0.015	94.2	14.3	14.2
0.39	0.016	89.6	14.8	14.6	0.45	0.016	88.4	13.4	13.3
0.39	0.017	84.3	13.9	13.8	0.45	0.017	83.2	12.7	12.5
0.40	0.010	143.0	23.3	23.2	0.46	0.010	141.0	21.2	21.1
0.40	0.011	130.0	21.1	21.1	0.46	0.011	128.2	19.2	19.2
0.40	0.012	119.2	19.4	19.3	0.46	0.012	117.5	17.6	17.6
0.40	0.013	110.0	17.9	17.8	0.46	0.013	108.5	16.3	16.2
0.40	0.014	102.2	16.6	16.5	0.46	0.014	100.7	15.1	15.0
0.40	0.015	95.4	15.5	15.4	0.46	0.015	94.0	14.1	14.0
0.40	0.016	89.4	14.5	14.4	0.46	0.016	88.1	13.2	13.1
0.40	0.017	84.1	13.7	13.6	0.46	0.017	82.9	12.5	12.3
0.41	0.010	142.7	22.9	22.8	0.47	0.010	140.6	20.8	20.8
0.41	0.011	129.7	20.8	20.7	0.47	0.011	127.9	18.9	18.9
0.41	0.012	118.9	19.1	19.0	0.47	0.012	117.2	17.4	17.3
0.41	0.013	109.8	17.6	17.5	0.47	0.013	108.2	16.0	15.9
0.41	0.014	101.9	16.4	16.3	0.47	0.014	100.5	14.9	14.8
0.41	0.015	95.1	15.3	15.2	0.47	0.015	93.8	13.9	13.8
0.41	0.016	89.2	14.3	14.2	0.47	0.016	87.9	13.0	12.9
0.41	0.017	84.0	13.5	13.3	0.47	0.017	82.7	12.3	12.1
0.42	0.010	142.4	22.5	22.5	0.48	0.010	140.3	20.5	20.4
0.42	0.011	129.5	20.5	20.4	0.48	0.011	127.5	18.6	18.6
0.42	0.012	118.7	18.8	18.7	0.48	0.012	116.9	17.1	17.0
0.42	0.013	109.5	17.3	17.3	0.48	0.013	107.9	15.8	15.7
0.42	0.014	101.7	16.1	16.0	0.48	0.014	100.2	14.6	14.5
0.42	0.015	94.9	15.0	14.9	0.48	0.015	93.5	13.7	13.6
0.42	0.016	89.0	14.1	14.0	0.48	0.016	87.7	12.8	12.7
0.42	0.017	83.8	13.3	13.1	0.48	0.017	82.5	12.1	11.9
0.43	0.010	142.1	22.2	22.1	0.49	0.010	139.9	20.2	20.1
0.43	0.011	129.1	20.2	20.1	0.49	0.011	127.2	18.3	18.3
0.43	0.012	118.4	18.5	18.4	0.49	0.012	116.6	16.8	16.7
0.43	0.013	109.3	17.1	17.0	0.49	0.013	107.6	15.5	15.4
0.43	0.014	101.5	15.9	15.8	0.49	0.014	99.9	14.4	14.3
0.43	0.015	94.7	14.8	14.7	0.49	0.015	93.3	13.4	13.3
0.43	0.016	88.8	13.9	13.8	0.49	0.016	87.4	12.6	12.5
0.43	0.017	83.6	13.1	12.9	0.49	0.017	82.3	11.9	11.7

Selecting the Perfect Spring for Your Barreled Spring Clock

Chart A-2.00: Only For Barrels With An Inside Diameter Of 2.00 Inches

(Arbor diameters of from 0.26 Inches thru 0.37 Inches, and spring thicknesses of from 0.010 Thru 0.017 Inches)

For An Arbor Diameter Of	And A Spring Thickness Of	The BEST Spring Length Is	Max Turns Using BEST Length	Max Turns If Spring Is 10" Short Or Long	For An Arbor Diameter Of	And A Spring Thickness Of	The BEST Spring Length Is	Max Turns Using BEST Length	Max Turns If Spring Is 10" Short Or Long
0.26	0.010	154.4	29.6	29.5	0.32	0.010	153.1	27.2	27.2
0.26	0.011	140.4	26.9	26.8	0.32	0.011	139.1	24.7	24.7
0.26	0.012	128.7	24.7	24.6	0.32	0.012	127.5	22.7	22.6
0.26	0.013	118.8	22.8	22.7	0.32	0.013	117.7	20.9	20.8
0.26	0.014	110.3	21.2	21.1	0.32	0.014	109.3	19.4	19.3
0.26	0.015	102.9	19.7	19.6	0.32	0.015	102.0	18.1	18.0
0.26	0.016	96.5	18.5	18.4	0.32	0.016	95.7	17.0	16.9
0.26	0.017	90.8	17.4	17.3	0.32	0.017	90.0	16.0	15.9
0.27	0.010	154.2	29.2	29.1	0.33	0.010	152.8	26.8	26.8
0.27	0.011	140.2	26.5	26.5	0.33	0.011	138.9	24.4	24.3
0.27	0.012	128.5	24.3	24.3	0.33	0.012	127.3	22.4	22.3
0.27	0.013	118.6	22.5	22.4	0.33	0.013	117.5	20.6	20.6
0.27	0.014	110.2	20.9	20.8	0.33	0.014	109.1	19.2	19.1
0.27	0.015	102.8	19.5	19.4	0.33	0.015	101.9	17.9	17.8
0.27	0.016	96.4	18.3	18.1	0.33	0.016	95.5	16.8	16.7
0.27	0.017	90.7	17.2	17.1	0.33	0.017	89.9	15.8	15.7
0.28	0.010	154.0	28.8	28.7	0.34	0.010	152.5	26.5	26.4
0.28	0.011	140.0	26.2	26.1	0.34	0.011	138.7	24.0	24.0
0.28	0.012	128.3	24.0	23.9	0.34	0.012	127.1	22.0	22.0
0.28	0.013	118.5	22.2	22.1	0.34	0.013	117.3	20.3	20.3
0.28	0.014	110.0	20.6	20.5	0.34	0.014	109.0	18.9	18.8
0.28	0.015	102.7	19.2	19.1	0.34	0.015	101.7	17.6	17.5
0.28	0.016	96.3	18.0	17.9	0.34	0.016	95.3	16.5	16.4
0.28	0.017	90.6	16.9	16.8	0.34	0.017	89.7	15.6	15.4
0.29	0.010	153.8	28.4	28.3	0.35	0.010	152.3	26.1	26.0
0.29	0.011	139.8	25.8	25.7	0.35	0.011	138.4	23.7	23.6
0.29	0.012	128.1	23.7	23.6	0.35	0.012	126.9	21.7	21.6
0.29	0.013	118.3	21.8	21.8	0.35	0.013	117.1	20.1	20.0
0.29	0.014	109.8	20.3	20.2	0.35	0.014	108.8	18.6	18.5
0.29	0.015	102.5	18.9	18.8	0.35	0.015	101.5	17.4	17.3
0.29	0.016	96.1	17.8	17.6	0.35	0.016	95.2	16.3	16.2
0.29	0.017	90.5	16.7	16.6	0.35	0.017	89.6	15.3	15.2
0.30	0.010	153.5	28.0	27.9	0.36	0.010	152.0	25.7	25.6
0.30	0.011	139.6	25.5	25.4	0.36	0.011	138.2	23.4	23.3
0.30	0.012	128.0	23.3	23.3	0.36	0.012	126.7	21.4	21.3
0.30	0.013	118.1	21.5	21.5	0.36	0.013	116.9	19.8	19.7
0.30	0.014	109.7	20.0	19.9	0.36	0.014	108.6	18.4	18.3
0.30	0.015	102.4	18.7	18.6	0.36	0.015	101.3	17.1	17.0
0.30	0.016	96.0	17.5	17.4	0.36	0.016	95.0	16.1	15.9
0.30	0.017	90.3	16.5	16.4	0.36	0.017	89.4	15.1	15.0
0.31	0.010	153.3	27.6	27.5	0.37	0.010	151.7	25.3	25.3
0.31	0.011	139.4	25.1	25.0	0.37	0.011	137.9	23.0	22.9
0.31	0.012	127.8	23.0	22.9	0.37	0.012	126.4	21.1	21.0
0.31	0.013	117.9	21.2	21.1	0.37	0.013	116.7	19.5	19.4
0.31	0.014	109.5	19.7	19.6	0.37	0.014	108.4	18.1	18.0
0.31	0.015	102.2	18.4	18.3	0.37	0.015	101.1	16.9	16.8
0.31	0.016	95.8	17.3	17.1	0.37	0.016	94.8	15.8	15.7
0.31	0.017	90.2	16.2	16.1	0.37	0.017	89.2	14.9	14.8

56

Chart A-2.00 (Continued): Only For Barrels With An Inside Diameter Of 2.00 Inches

(Arbor diameters of from 0.38 Inches thru 0.49 Inches, and spring thicknesses of from 0.010 Thru 0.017 Inches)

For An Arbor Diameter Of	And A Spring Thickness Of	The BEST Spring Length Is	Max Turns Using BEST Length	Max Turns If Spring Is 10" Short Or Long	For An Arbor Diameter Of	And A Spring Thickness Of	The BEST Spring Length Is	Max Turns Using BEST Length	Max Turns If Spring Is 10" Short Or Long
0.38	0.010	151.4	25.0	24.9	0.44	0.010	149.5	22.8	22.7
0.38	0.011	137.6	22.7	22.6	0.44	0.011	135.9	20.7	20.7
0.38	0.012	126.2	20.8	20.7	0.44	0.012	124.6	19.0	18.9
0.38	0.013	116.5	19.2	19.1	0.44	0.013	115.0	17.5	17.5
0.38	0.014	108.1	17.8	17.7	0.44	0.014	106.8	16.3	16.2
0.38	0.015	100.9	16.6	16.5	0.44	0.015	99.7	15.2	15.1
0.38	0.016	94.6	15.6	15.5	0.44	0.016	93.4	14.3	14.1
0.38	0.017	89.1	14.7	14.6	0.44	0.017	87.9	13.4	13.3
0.39	0.010	151.1	24.6	24.5	0.45	0.010	149.1	22.5	22.4
0.39	0.011	137.4	22.4	22.3	0.45	0.011	135.6	20.4	20.3
0.39	0.012	125.9	20.5	20.4	0.45	0.012	124.3	18.7	18.6
0.39	0.013	116.2	18.9	18.8	0.45	0.013	114.7	17.3	17.2
0.39	0.014	107.9	17.6	17.5	0.45	0.014	106.5	16.0	15.9
0.39	0.015	100.7	16.4	16.3	0.45	0.015	99.4	15.0	14.9
0.39	0.016	94.4	15.4	15.3	0.45	0.016	93.2	14.0	13.9
0.39	0.017	88.9	14.5	14.3	0.45	0.017	87.7	13.2	13.1
0.40	0.010	150.8	24.2	24.2	0.46	0.010	148.8	22.1	22.0
0.40	0.011	137.1	22.0	21.9	0.46	0.011	135.2	20.1	20.0
0.40	0.012	125.7	20.2	20.1	0.46	0.012	124.0	18.4	18.3
0.40	0.013	116.0	18.6	18.5	0.46	0.013	114.4	17.0	16.9
0.40	0.014	107.7	17.3	17.2	0.46	0.014	106.3	15.8	15.7
0.40	0.015	100.5	16.1	16.0	0.46	0.015	99.2	14.7	14.6
0.40	0.016	94.2	15.1	15.0	0.46	0.016	93.0	13.8	13.7
0.40	0.017	88.7	14.2	14.1	0.46	0.017	87.5	13.0	12.9
0.41	0.010	150.5	23.9	23.8	0.47	0.010	148.4	21.8	21.7
0.41	0.011	136.8	21.7	21.6	0.47	0.011	134.9	19.8	19.7
0.41	0.012	125.4	19.9	19.8	0.47	0.012	123.7	18.1	18.1
0.41	0.013	115.8	18.4	18.3	0.47	0.013	114.2	16.7	16.7
0.41	0.014	107.5	17.0	17.0	0.47	0.014	106.0	15.6	15.5
0.41	0.015	100.3	15.9	15.8	0.47	0.015	98.9	14.5	14.4
0.41	0.016	94.0	14.9	14.8	0.47	0.016	92.8	13.6	13.5
0.41	0.017	88.5	14.0	13.9	0.47	0.017	87.3	12.8	12.7
0.42	0.010	150.2	23.5	23.4	0.48	0.010	148.0	21.4	21.4
0.42	0.011	136.5	21.4	21.3	0.48	0.011	134.6	19.5	19.4
0.42	0.012	125.1	19.6	19.5	0.48	0.012	123.4	17.9	17.8
0.42	0.013	115.5	18.1	18.0	0.48	0.013	113.9	16.5	16.4
0.42	0.014	107.3	16.8	16.7	0.48	0.014	105.7	15.3	15.2
0.42	0.015	100.1	15.7	15.6	0.48	0.015	98.7	14.3	14.2
0.42	0.016	93.8	14.7	14.6	0.48	0.016	92.5	13.4	13.3
0.42	0.017	88.3	13.8	13.7	0.48	0.017	87.1	12.6	12.5
0.43	0.010	149.8	23.2	23.1	0.49	0.010	147.7	21.1	21.0
0.43	0.011	136.2	21.0	21.0	0.49	0.011	134.2	19.2	19.1
0.43	0.012	124.8	19.3	19.2	0.49	0.012	123.0	17.6	17.5
0.43	0.013	115.2	17.8	17.7	0.49	0.013	113.6	16.2	16.1
0.43	0.014	107.0	16.5	16.4	0.49	0.014	105.5	15.1	15.0
0.43	0.015	99.9	15.4	15.3	0.49	0.015	98.4	14.1	14.0
0.43	0.016	93.6	14.5	14.4	0.49	0.016	92.3	13.2	13.1
0.43	0.017	88.1	13.6	13.5	0.49	0.017	86.9	12.4	12.3

Chart A-2.10: Only For Barrels With An Inside Diameter Of 2.10 Inches

(Arbor diameters of from 0.26 Inches thru 0.37 Inches, and spring thicknesses of from 0.010 Thru 0.017 Inches)

For An Arbor Diameter Of	And A Spring Thickness Of	The BEST Spring Length Is	Max Turns Using BEST Length	Max Turns If Spring Is 10" Short Or Long	For An Arbor Diameter Of	And A Spring Thickness Of	The BEST Spring Length Is	Max Turns Using BEST Length	Max Turns If Spring Is 10" Short Or Long
0.26	0.010	170.5	31.6	31.6	0.32	0.010	169.2	29.2	29.1
0.26	0.011	155.0	28.8	28.7	0.32	0.011	153.8	26.6	26.5
0.26	0.012	142.1	26.4	26.3	0.32	0.012	141.0	24.3	24.3
0.26	0.013	131.2	24.3	24.2	0.32	0.013	130.1	22.5	22.4
0.26	0.014	121.8	22.6	22.5	0.32	0.014	120.8	20.9	20.8
0.26	0.015	113.7	21.1	21.0	0.32	0.015	112.8	19.5	19.4
0.26	0.016	106.6	19.8	19.7	0.32	0.016	105.7	18.3	18.2
0.26	0.017	100.3	18.6	18.5	0.32	0.017	99.5	17.2	17.1
0.27	0.010	170.3	31.2	31.2	0.33	0.010	168.9	28.8	28.8
0.27	0.011	154.8	28.4	28.3	0.33	0.011	153.5	26.2	26.1
0.27	0.012	141.9	26.0	25.9	0.33	0.012	140.8	24.0	23.9
0.27	0.013	131.0	24.0	23.9	0.33	0.013	129.9	22.2	22.1
0.27	0.014	121.7	22.3	22.2	0.33	0.014	120.6	20.6	20.5
0.27	0.015	113.5	20.8	20.7	0.33	0.015	112.6	19.2	19.1
0.27	0.016	106.4	19.5	19.4	0.33	0.016	105.6	18.0	17.9
0.27	0.017	100.2	18.4	18.3	0.33	0.017	99.4	16.9	16.8
0.28	0.010	170.1	30.8	30.7	0.34	0.010	168.6	28.4	28.4
0.28	0.011	154.6	28.0	27.9	0.34	0.011	153.3	25.8	25.8
0.28	0.012	141.8	25.7	25.6	0.34	0.012	140.5	23.7	23.6
0.28	0.013	130.8	23.7	23.6	0.34	0.013	129.7	21.9	21.8
0.28	0.014	121.5	22.0	21.9	0.34	0.014	120.5	20.3	20.2
0.28	0.015	113.4	20.5	20.4	0.34	0.015	112.4	19.0	18.9
0.28	0.016	106.3	19.3	19.2	0.34	0.016	105.4	17.8	17.7
0.28	0.017	100.1	18.1	18.0	0.34	0.017	99.2	16.7	16.6
0.29	0.010	169.9	30.4	30.3	0.35	0.010	168.4	28.0	28.0
0.29	0.011	154.4	27.6	27.6	0.35	0.011	153.1	25.5	25.4
0.29	0.012	141.6	25.3	25.3	0.35	0.012	140.3	23.4	23.3
0.29	0.013	130.7	23.4	23.3	0.35	0.013	129.5	21.6	21.5
0.29	0.014	121.3	21.7	21.6	0.35	0.014	120.3	20.0	19.9
0.29	0.015	113.3	20.3	20.2	0.35	0.015	112.2	18.7	18.6
0.29	0.016	106.2	19.0	18.9	0.35	0.016	105.2	17.5	17.4
0.29	0.017	99.9	17.9	17.8	0.35	0.017	99.0	16.5	16.4
0.30	0.010	169.6	30.0	29.9	0.36	0.010	168.1	27.7	27.6
0.30	0.011	154.2	27.3	27.2	0.36	0.011	152.8	25.1	25.1
0.30	0.012	141.4	25.0	24.9	0.36	0.012	140.1	23.0	23.0
0.30	0.013	130.5	23.1	23.0	0.36	0.013	129.3	21.3	21.2
0.30	0.014	121.2	21.4	21.3	0.36	0.014	120.1	19.8	19.7
0.30	0.015	113.1	20.0	19.9	0.36	0.015	112.1	18.4	18.3
0.30	0.016	106.0	18.8	18.7	0.36	0.016	105.1	17.3	17.2
0.30	0.017	99.8	17.6	17.5	0.36	0.017	98.9	16.3	16.2
0.31	0.010	169.4	29.6	29.5	0.37	0.010	167.8	27.3	27.2
0.31	0.011	154.0	26.9	26.8	0.37	0.011	152.5	24.8	24.7
0.31	0.012	141.2	24.7	24.6	0.37	0.012	139.8	22.7	22.7
0.31	0.013	130.3	22.8	22.7	0.37	0.013	129.1	21.0	20.9
0.31	0.014	121.0	21.1	21.1	0.37	0.014	119.9	19.5	19.4
0.31	0.015	112.9	19.7	19.6	0.37	0.015	111.9	18.2	18.1
0.31	0.016	105.9	18.5	18.4	0.37	0.016	104.9	17.0	17.0
0.31	0.017	99.7	17.4	17.3	0.37	0.017	98.7	16.0	15.9

Chart A-2.10 (Continued): Only For Barrels With An Inside Diameter Of 2.10 Inches

(Arbor diameters of from 0.38 Inches thru 0.49 Inches, and spring thicknesses of from 0.010 Thru 0.017 Inches)

For An Arbor Diameter Of	And A Spring Thickness Of	The BEST Spring Length Is	Max Turns Using BEST Length	Max Turns If Spring Is 10" Short Or Long	For An Arbor Diameter Of	And A Spring Thickness Of	The BEST Spring Length Is	Max Turns Using BEST Length	Max Turns If Spring Is 10" Short Or Long
0.38	0.010	167.5	26.9	26.8	0.44	0.010	165.6	24.7	24.7
0.38	0.011	152.3	24.5	24.4	0.44	0.011	150.5	22.5	22.4
0.38	0.012	139.6	22.4	22.3	0.44	0.012	138.0	20.6	20.5
0.38	0.013	128.9	20.7	20.6	0.44	0.013	127.4	19.0	18.9
0.38	0.014	119.6	19.2	19.1	0.44	0.014	118.3	17.7	17.6
0.38	0.015	111.7	17.9	17.8	0.44	0.015	110.4	16.5	16.4
0.38	0.016	104.7	16.8	16.7	0.44	0.016	103.5	15.4	15.4
0.38	0.017	98.5	15.8	15.7	0.44	0.017	97.4	14.5	14.4
0.39	0.010	167.2	26.5	26.5	0.45	0.010	165.2	24.4	24.3
0.39	0.011	152.0	24.1	24.1	0.45	0.011	150.2	22.1	22.1
0.39	0.012	139.3	22.1	22.0	0.45	0.012	137.7	20.3	20.2
0.39	0.013	128.6	20.4	20.3	0.45	0.013	127.1	18.7	18.7
0.39	0.014	119.4	19.0	18.9	0.45	0.014	118.0	17.4	17.3
0.39	0.015	111.5	17.7	17.6	0.45	0.015	110.2	16.2	16.2
0.39	0.016	104.5	16.6	16.5	0.45	0.016	103.3	15.2	15.1
0.39	0.017	98.4	15.6	15.5	0.45	0.017	97.2	14.3	14.2
0.40	0.010	166.9	26.2	26.1	0.46	0.010	164.9	24.0	24.0
0.40	0.011	151.7	23.8	23.7	0.46	0.011	149.9	21.8	21.8
0.40	0.012	139.1	21.8	21.7	0.46	0.012	137.4	20.0	19.9
0.40	0.013	128.4	20.1	20.0	0.46	0.013	126.8	18.5	18.4
0.40	0.014	119.2	18.7	18.6	0.46	0.014	117.8	17.2	17.1
0.40	0.015	111.3	17.4	17.4	0.46	0.015	109.9	16.0	15.9
0.40	0.016	104.3	16.4	16.3	0.46	0.016	103.0	15.0	14.9
0.40	0.017	98.2	15.4	15.3	0.46	0.017	97.0	14.1	14.0
0.41	0.010	166.6	25.8	25.7	0.47	0.010	164.5	23.7	23.6
0.41	0.011	151.4	23.5	23.4	0.47	0.011	149.6	21.5	21.5
0.41	0.012	138.8	21.5	21.4	0.47	0.012	137.1	19.7	19.7
0.41	0.013	128.1	19.8	19.8	0.47	0.013	126.5	18.2	18.1
0.41	0.014	119.0	18.4	18.3	0.47	0.014	117.5	16.9	16.8
0.41	0.015	111.1	17.2	17.1	0.47	0.015	109.7	15.8	15.7
0.41	0.016	104.1	16.1	16.0	0.47	0.016	102.8	14.8	14.7
0.41	0.017	98.0	15.2	15.1	0.47	0.017	96.8	13.9	13.8
0.42	0.010	166.3	25.4	25.4	0.48	0.010	164.1	23.3	23.3
0.42	0.011	151.1	23.1	23.1	0.48	0.011	149.2	21.2	21.1
0.42	0.012	138.5	21.2	21.1	0.48	0.012	136.8	19.4	19.4
0.42	0.013	127.9	19.6	19.5	0.48	0.013	126.3	17.9	17.9
0.42	0.014	118.8	18.2	18.1	0.48	0.014	117.2	16.7	16.6
0.42	0.015	110.8	17.0	16.9	0.48	0.015	109.4	15.5	15.5
0.42	0.016	103.9	15.9	15.8	0.48	0.016	102.6	14.6	14.5
0.42	0.017	97.8	15.0	14.9	0.48	0.017	96.5	13.7	13.6
0.43	0.010	165.9	25.1	25.0	0.49	0.010	163.8	23.0	22.9
0.43	0.011	150.8	22.8	22.7	0.49	0.011	148.9	20.9	20.8
0.43	0.012	138.3	20.9	20.8	0.49	0.012	136.5	19.2	19.1
0.43	0.013	127.6	19.3	19.2	0.49	0.013	126.0	17.7	17.6
0.43	0.014	118.5	17.9	17.8	0.49	0.014	117.0	16.4	16.3
0.43	0.015	110.6	16.7	16.6	0.49	0.015	109.2	15.3	15.2
0.43	0.016	103.7	15.7	15.6	0.49	0.016	102.3	14.4	14.3
0.43	0.017	97.6	14.7	14.6	0.49	0.017	96.3	13.5	13.4

Chart A-2.20: Only For Barrels With An Inside Diameter Of 2.20 Inches

(Arbor diameters of from 0.26 Inches thru 0.37 Inches, and spring thicknesses of from 0.010 Thru 0.017 Inches)

For An Arbor Diameter Of	And A Spring Thickness Of	The BEST Spring Length Is	Max Turns Using BEST Length	Max Turns If Spring Is 10" Short Or Long	For An Arbor Diameter Of	And A Spring Thickness Of	The BEST Spring Length Is	Max Turns Using BEST Length	Max Turns If Spring Is 10" Short Or Long
0.26	0.010	187.4	33.6	33.6	0.32	0.010	186.0	31.2	31.1
0.26	0.011	170.4	30.6	30.5	0.32	0.011	169.1	28.4	28.3
0.26	0.012	156.2	28.0	28.0	0.32	0.012	155.0	26.0	25.9
0.26	0.013	144.2	25.9	25.8	0.32	0.013	143.1	24.0	23.9
0.26	0.014	133.9	24.0	24.0	0.32	0.014	132.9	22.3	22.2
0.26	0.015	124.9	22.4	22.4	0.32	0.015	124.0	20.8	20.7
0.26	0.016	117.1	21.0	20.9	0.32	0.016	116.3	19.5	19.4
0.26	0.017	110.2	19.8	19.7	0.32	0.017	109.4	18.4	18.3
0.27	0.010	187.2	33.2	33.2	0.33	0.010	185.8	30.8	30.8
0.27	0.011	170.2	30.2	30.2	0.33	0.011	168.9	28.0	27.9
0.27	0.012	156.0	27.7	27.6	0.33	0.012	154.8	25.7	25.6
0.27	0.013	144.0	25.6	25.5	0.33	0.013	142.9	23.7	23.6
0.27	0.014	133.7	23.7	23.7	0.33	0.014	132.7	22.0	21.9
0.27	0.015	124.8	22.2	22.1	0.33	0.015	123.9	20.5	20.5
0.27	0.016	117.0	20.8	20.7	0.33	0.016	116.1	19.3	19.2
0.27	0.017	110.1	19.5	19.5	0.33	0.017	109.3	18.1	18.0
0.28	0.010	187.0	32.8	32.8	0.34	0.010	185.5	30.4	30.4
0.28	0.011	170.0	29.8	29.8	0.34	0.011	168.7	27.6	27.6
0.28	0.012	155.8	27.3	27.3	0.34	0.012	154.6	25.3	25.3
0.28	0.013	143.8	25.2	25.2	0.34	0.013	142.7	23.4	23.3
0.28	0.014	133.6	23.4	23.4	0.34	0.014	132.5	21.7	21.6
0.28	0.015	124.7	21.9	21.8	0.34	0.015	123.7	20.3	20.2
0.28	0.016	116.9	20.5	20.4	0.34	0.016	116.0	19.0	18.9
0.28	0.017	110.0	19.3	19.2	0.34	0.017	109.1	17.9	17.8
0.29	0.010	186.8	32.4	32.4	0.35	0.010	185.3	30.0	30.0
0.29	0.011	169.8	29.5	29.4	0.35	0.011	168.4	27.3	27.2
0.29	0.012	155.6	27.0	26.9	0.35	0.012	154.4	25.0	25.0
0.29	0.013	143.7	24.9	24.9	0.35	0.013	142.5	23.1	23.0
0.29	0.014	133.4	23.1	23.1	0.35	0.014	132.3	21.4	21.4
0.29	0.015	124.5	21.6	21.5	0.35	0.015	123.5	20.0	19.9
0.29	0.016	116.7	20.3	20.2	0.35	0.016	115.8	18.8	18.7
0.29	0.017	109.9	19.1	19.0	0.35	0.017	109.0	17.7	17.6
0.30	0.010	186.5	32.0	32.0	0.36	0.010	185.0	29.6	29.6
0.30	0.011	169.6	29.1	29.0	0.36	0.011	168.2	26.9	26.9
0.30	0.012	155.4	26.7	26.6	0.36	0.012	154.1	24.7	24.6
0.30	0.013	143.5	24.6	24.5	0.36	0.013	142.3	22.8	22.7
0.30	0.014	133.2	22.9	22.8	0.36	0.014	132.1	21.2	21.1
0.30	0.015	124.4	21.3	21.3	0.36	0.015	123.3	19.8	19.7
0.30	0.016	116.6	20.0	19.9	0.36	0.016	115.6	18.5	18.4
0.30	0.017	109.7	18.8	18.7	0.36	0.017	108.8	17.4	17.3
0.31	0.010	186.3	31.6	31.5	0.37	0.010	184.7	29.2	29.2
0.31	0.011	169.4	28.7	28.7	0.37	0.011	167.9	26.6	26.5
0.31	0.012	155.2	26.3	26.3	0.37	0.012	153.9	24.4	24.3
0.31	0.013	143.3	24.3	24.2	0.37	0.013	142.1	22.5	22.4
0.31	0.014	133.1	22.6	22.5	0.37	0.014	131.9	20.9	20.8
0.31	0.015	124.2	21.1	21.0	0.37	0.015	123.1	19.5	19.4
0.31	0.016	116.4	19.8	19.7	0.37	0.016	115.4	18.3	18.2
0.31	0.017	109.6	18.6	18.5	0.37	0.017	108.6	17.2	17.1

Chart A-2.20 (Continued): Only For Barrels With An Inside Diameter Of 2.20 Inches

(Arbor diameters of from 0.38 Inches thru 0.49 Inches, and spring thicknesses of from 0.010 Thru 0.017 Inches)

For An Arbor Diameter Of	And A Spring Thickness Of	The BEST Spring Length Is	Max Turns Using BEST Length	Max Turns If Spring Is 10" Short Or Long	For An Arbor Diameter Of	And A Spring Thickness Of	The BEST Spring Length Is	Max Turns Using BEST Length	Max Turns If Spring Is 10" Short Or Long
0.38	0.010	184.4	28.9	28.8	0.44	0.010	182.5	26.6	26.6
0.38	0.011	167.6	26.2	26.2	0.44	0.011	165.9	24.2	24.2
0.38	0.012	153.7	24.1	24.0	0.44	0.012	152.1	22.2	22.1
0.38	0.013	141.8	22.2	22.1	0.44	0.013	140.4	20.5	20.4
0.38	0.014	131.7	20.6	20.5	0.44	0.014	130.3	19.0	19.0
0.38	0.015	122.9	19.2	19.2	0.44	0.015	121.6	17.8	17.7
0.38	0.016	115.2	18.0	18.0	0.44	0.016	114.0	16.7	16.6
0.38	0.017	108.5	17.0	16.9	0.44	0.017	107.3	15.7	15.6
0.39	0.010	184.1	28.5	28.4	0.45	0.010	182.1	26.3	26.2
0.39	0.011	167.4	25.9	25.8	0.45	0.011	165.6	23.9	23.8
0.39	0.012	153.4	23.7	23.7	0.45	0.012	151.8	21.9	21.8
0.39	0.013	141.6	21.9	21.8	0.45	0.013	140.1	20.2	20.2
0.39	0.014	131.5	20.3	20.3	0.45	0.014	130.1	18.8	18.7
0.39	0.015	122.7	19.0	18.9	0.45	0.015	121.4	17.5	17.4
0.39	0.016	115.1	17.8	17.7	0.45	0.016	113.8	16.4	16.3
0.39	0.017	108.3	16.8	16.7	0.45	0.017	107.1	15.5	15.4
0.40	0.010	183.8	28.1	28.1	0.46	0.010	181.8	25.9	25.9
0.40	0.011	167.1	25.6	25.5	0.46	0.011	165.2	23.6	23.5
0.40	0.012	153.2	23.4	23.4	0.46	0.012	151.5	21.6	21.5
0.40	0.013	141.4	21.6	21.6	0.46	0.013	139.8	19.9	19.9
0.40	0.014	131.3	20.1	20.0	0.46	0.014	129.8	18.5	18.4
0.40	0.015	122.5	18.7	18.7	0.46	0.015	121.2	17.3	17.2
0.40	0.016	114.9	17.6	17.5	0.46	0.016	113.6	16.2	16.1
0.40	0.017	108.1	16.5	16.5	0.46	0.017	106.9	15.3	15.2
0.41	0.010	183.5	27.7	27.7	0.47	0.010	181.4	25.6	25.5
0.41	0.011	166.8	25.2	25.2	0.47	0.011	164.9	23.2	23.2
0.41	0.012	152.9	23.1	23.1	0.47	0.012	151.2	21.3	21.3
0.41	0.013	141.1	21.3	21.3	0.47	0.013	139.5	19.7	19.6
0.41	0.014	131.0	19.8	19.7	0.47	0.014	129.6	18.3	18.2
0.41	0.015	122.3	18.5	18.4	0.47	0.015	120.9	17.0	17.0
0.41	0.016	114.7	17.3	17.3	0.47	0.016	113.4	16.0	15.9
0.41	0.017	107.9	16.3	16.2	0.47	0.017	106.7	15.0	15.0
0.42	0.010	183.1	27.4	27.3	0.48	0.010	181.0	25.2	25.2
0.42	0.011	166.5	24.9	24.8	0.48	0.011	164.6	22.9	22.9
0.42	0.012	152.6	22.8	22.7	0.48	0.012	150.8	21.0	21.0
0.42	0.013	140.9	21.1	21.0	0.48	0.013	139.2	19.4	19.3
0.42	0.014	130.8	19.6	19.5	0.48	0.014	129.3	18.0	17.9
0.42	0.015	122.1	18.2	18.2	0.48	0.015	120.7	16.8	16.7
0.42	0.016	114.5	17.1	17.0	0.48	0.016	113.1	15.8	15.7
0.42	0.017	107.7	16.1	16.0	0.48	0.017	106.5	14.8	14.8
0.43	0.010	182.8	27.0	27.0	0.49	0.010	180.6	24.9	24.8
0.43	0.011	166.2	24.6	24.5	0.49	0.011	164.2	22.6	22.6
0.43	0.012	152.3	22.5	22.4	0.49	0.012	150.5	20.7	20.7
0.43	0.013	140.6	20.8	20.7	0.49	0.013	139.0	19.1	19.1
0.43	0.014	130.6	19.3	19.2	0.49	0.014	129.0	17.8	17.7
0.43	0.015	121.9	18.0	17.9	0.49	0.015	120.4	16.6	16.5
0.43	0.016	114.3	16.9	16.8	0.49	0.016	112.9	15.5	15.5
0.43	0.017	107.5	15.9	15.8	0.49	0.017	106.3	14.6	14.5

Selecting the Perfect Spring for Your Barreled Spring Clock

Chart A-2.40: Only For Barrels With An Inside Diameter Of 2.40 Inches

(Arbor diameters of from 0.26 Inches thru 0.37 Inches, and spring thicknesses of from 0.010 Thru 0.017 Inches)

For An Arbor Diameter Of	And A Spring Thickness Of	The BEST Spring Length Is	Max Turns Using BEST Length	Max Turns If Spring Is 10" Short Or Long	For An Arbor Diameter Of	And A Spring Thickness Of	The BEST Spring Length Is	Max Turns Using BEST Length	Max Turns If Spring Is 10" Short Or Long
0.26	0.010	223.5	37.7	37.7	0.32	0.010	222.2	35.2	35.2
0.26	0.011	203.2	34.3	34.2	0.32	0.011	202.0	32.0	32.0
0.26	0.012	186.3	31.4	31.4	0.32	0.012	185.1	29.3	29.3
0.26	0.013	172.0	29.0	28.9	0.32	0.013	170.9	27.1	27.0
0.26	0.014	159.7	26.9	26.9	0.32	0.014	158.7	25.1	25.1
0.26	0.015	149.0	25.1	25.1	0.32	0.015	148.1	23.5	23.4
0.26	0.016	139.7	23.6	23.5	0.32	0.016	138.9	22.0	21.9
0.26	0.017	131.5	22.2	22.1	0.32	0.017	130.7	20.7	20.6
0.27	0.010	223.3	37.3	37.2	0.33	0.010	221.9	34.8	34.8
0.27	0.011	203.0	33.9	33.8	0.33	0.011	201.7	31.6	31.6
0.27	0.012	186.1	31.1	31.0	0.33	0.012	184.9	29.0	29.0
0.27	0.013	171.8	28.7	28.6	0.33	0.013	170.7	26.8	26.7
0.27	0.014	159.5	26.6	26.6	0.33	0.014	158.5	24.9	24.8
0.27	0.015	148.9	24.9	24.8	0.33	0.015	147.9	23.2	23.1
0.27	0.016	139.6	23.3	23.2	0.33	0.016	138.7	21.8	21.7
0.27	0.017	131.4	21.9	21.9	0.33	0.017	130.5	20.5	20.4
0.28	0.010	223.1	36.9	36.8	0.34	0.010	221.7	34.4	34.4
0.28	0.011	202.8	33.5	33.5	0.34	0.011	201.5	31.3	31.2
0.28	0.012	185.9	30.7	30.7	0.34	0.012	184.7	28.7	28.6
0.28	0.013	171.6	28.4	28.3	0.34	0.013	170.5	26.5	26.4
0.28	0.014	159.4	26.3	26.3	0.34	0.014	158.3	24.6	24.5
0.28	0.015	148.7	24.6	24.5	0.34	0.015	147.8	22.9	22.9
0.28	0.016	139.4	23.0	23.0	0.34	0.016	138.5	21.5	21.4
0.28	0.017	131.2	21.7	21.6	0.34	0.017	130.4	20.2	20.2
0.29	0.010	222.9	36.4	36.4	0.35	0.010	221.4	34.0	34.0
0.29	0.011	202.6	33.1	33.1	0.35	0.011	201.3	30.9	30.9
0.29	0.012	185.7	30.4	30.3	0.35	0.012	184.5	28.3	28.3
0.29	0.013	171.5	28.0	28.0	0.35	0.013	170.3	26.2	26.1
0.29	0.014	159.2	26.0	26.0	0.35	0.014	158.1	24.3	24.2
0.29	0.015	148.6	24.3	24.2	0.35	0.015	147.6	22.7	22.6
0.29	0.016	139.3	22.8	22.7	0.35	0.016	138.4	21.3	21.2
0.29	0.017	131.1	21.4	21.4	0.35	0.017	130.2	20.0	19.9
0.30	0.010	222.7	36.0	36.0	0.36	0.010	221.1	33.6	33.6
0.30	0.011	202.4	32.8	32.7	0.36	0.011	201.0	30.5	30.5
0.30	0.012	185.6	30.0	30.0	0.36	0.012	184.3	28.0	28.0
0.30	0.013	171.3	27.7	27.7	0.36	0.013	170.1	25.8	25.8
0.30	0.014	159.0	25.7	25.7	0.36	0.014	157.9	24.0	23.9
0.30	0.015	148.4	24.0	24.0	0.36	0.015	147.4	22.4	22.3
0.30	0.016	139.2	22.5	22.5	0.36	0.016	138.2	21.0	20.9
0.30	0.017	131.0	21.2	21.1	0.36	0.017	130.1	19.8	19.7
0.31	0.010	222.4	35.6	35.6	0.37	0.010	220.8	33.2	33.2
0.31	0.011	202.2	32.4	32.3	0.37	0.011	200.7	30.2	30.1
0.31	0.012	185.4	29.7	29.6	0.37	0.012	184.0	27.7	27.6
0.31	0.013	171.1	27.4	27.3	0.37	0.013	169.9	25.5	25.5
0.31	0.014	158.9	25.4	25.4	0.37	0.014	157.7	23.7	23.7
0.31	0.015	148.3	23.7	23.7	0.37	0.015	147.2	22.1	22.1
0.31	0.016	139.0	22.3	22.2	0.37	0.016	138.0	20.8	20.7
0.31	0.017	130.8	21.0	20.9	0.37	0.017	129.9	19.5	19.5

Chart A-2.40 (Continued): Only For Barrels With An Inside Diameter Of 2.40 Inches

(Arbor diameters of from 0.38 Inches thru 0.49 Inches, and spring thicknesses of from 0.010 Thru 0.017 Inches)

For An Arbor Diameter Of	And A Spring Thickness Of	The BEST Spring Length Is	Max Turns Using BEST Length	Max Turns If Spring Is 10" Short Or Long	For An Arbor Diameter Of	And A Spring Thickness Of	The BEST Spring Length Is	Max Turns Using BEST Length	Max Turns If Spring Is 10" Short Or Long
0.38	0.010	220.5	32.8	32.8	0.44	0.010	218.6	30.5	30.5
0.38	0.011	200.5	29.8	29.8	0.44	0.011	198.7	27.8	27.7
0.38	0.012	183.8	27.3	27.3	0.44	0.012	182.2	25.4	25.4
0.38	0.013	169.6	25.2	25.2	0.44	0.013	168.1	23.5	23.4
0.38	0.014	157.5	23.4	23.4	0.44	0.014	156.1	21.8	21.8
0.38	0.015	147.0	21.9	21.8	0.44	0.015	145.7	20.4	20.3
0.38	0.016	137.8	20.5	20.4	0.44	0.016	136.6	19.1	19.0
0.38	0.017	129.7	19.3	19.2	0.44	0.017	128.6	18.0	17.9
0.39	0.010	220.2	32.4	32.4	0.45	0.010	218.2	30.2	30.1
0.39	0.011	200.2	29.5	29.4	0.45	0.011	198.4	27.4	27.4
0.39	0.012	183.5	27.0	27.0	0.45	0.012	181.9	25.1	25.1
0.39	0.013	169.4	24.9	24.9	0.45	0.013	167.9	23.2	23.2
0.39	0.014	157.3	23.2	23.1	0.45	0.014	155.9	21.5	21.5
0.39	0.015	146.8	21.6	21.6	0.45	0.015	145.5	20.1	20.0
0.39	0.016	137.6	20.3	20.2	0.45	0.016	136.4	18.9	18.8
0.39	0.017	129.5	19.1	19.0	0.45	0.017	128.4	17.7	17.7
0.40	0.010	219.9	32.0	32.0	0.46	0.010	217.9	29.8	29.8
0.40	0.011	199.9	29.1	29.1	0.46	0.011	198.1	27.1	27.0
0.40	0.012	183.3	26.7	26.7	0.46	0.012	181.6	24.8	24.8
0.40	0.013	169.2	24.7	24.6	0.46	0.013	167.6	22.9	22.9
0.40	0.014	157.1	22.9	22.8	0.46	0.014	155.6	21.3	21.2
0.40	0.015	146.6	21.4	21.3	0.46	0.015	145.3	19.9	19.8
0.40	0.016	137.4	20.0	20.0	0.46	0.016	136.2	18.6	18.6
0.40	0.017	129.4	18.9	18.8	0.46	0.017	128.2	17.5	17.5
0.41	0.010	219.6	31.7	31.6	0.47	0.010	217.5	29.4	29.4
0.41	0.011	199.6	28.8	28.7	0.47	0.011	197.7	26.8	26.7
0.41	0.012	183.0	26.4	26.3	0.47	0.012	181.3	24.5	24.5
0.41	0.013	168.9	24.4	24.3	0.47	0.013	167.3	22.6	22.6
0.41	0.014	156.9	22.6	22.6	0.47	0.014	155.4	21.0	21.0
0.41	0.015	146.4	21.1	21.0	0.47	0.015	145.0	19.6	19.6
0.41	0.016	137.2	19.8	19.7	0.47	0.016	135.9	18.4	18.3
0.41	0.017	129.2	18.6	18.6	0.47	0.017	128.0	17.3	17.2
0.42	0.010	219.3	31.3	31.2	0.48	0.010	217.1	29.1	29.0
0.42	0.011	199.3	28.4	28.4	0.48	0.011	197.4	26.4	26.4
0.42	0.012	182.7	26.1	26.0	0.48	0.012	181.0	24.2	24.2
0.42	0.013	168.7	24.1	24.0	0.48	0.013	167.0	22.4	22.3
0.42	0.014	156.6	22.3	22.3	0.48	0.014	155.1	20.8	20.7
0.42	0.015	146.2	20.9	20.8	0.48	0.015	144.8	19.4	19.3
0.42	0.016	137.0	19.6	19.5	0.48	0.016	135.7	18.2	18.1
0.42	0.017	129.0	18.4	18.3	0.48	0.017	127.7	17.1	17.0
0.43	0.010	218.9	30.9	30.9	0.49	0.010	216.8	28.7	28.7
0.43	0.011	199.0	28.1	28.1	0.49	0.011	197.1	26.1	26.1
0.43	0.012	182.4	25.8	25.7	0.49	0.012	180.6	23.9	23.9
0.43	0.013	168.4	23.8	23.7	0.49	0.013	166.7	22.1	22.0
0.43	0.014	156.4	22.1	22.0	0.49	0.014	154.8	20.5	20.5
0.43	0.015	146.0	20.6	20.5	0.49	0.015	144.5	19.1	19.1
0.43	0.016	136.8	19.3	19.3	0.49	0.016	135.5	17.9	17.9
0.43	0.017	128.8	18.2	18.1	0.49	0.017	127.5	16.9	16.8

Awe struck by a basic clock.

Finding Beats Per Hour and Pendulum Length
(WITHOUT TAKING THE MOVEMENT PLATES APART)

Below is a method to determine the beats per hour (BPH) and pendulum length for any movement you may have on your bench (WITHOUT TAKING THE MOVEMENT PLATES APART).

1. Let down the springs.
2. Move the verge away from the escape wheel, so the escape wheel can turn freely.
3. Mark a tooth on the escape wheel so that you can watch the number of turns the escape wheel makes.
4. Mount a minute hand on the movement.
5. Operate the wheel train with the minute hand so that the minute hand moves through 30 minutes of action. Count the number of times the escape wheel turns.
6. Count the number of teeth in the escape wheel.
7. The beats per hour equals the number of escape wheel teeth times the number of revolutions of the escape wheel makes times 4.

BPH = number of escape teeth X number of escape wheel revolutions X 4

Now that you have the BPH, you can use Chart B (on the following page) to find the pendulum length.

Chart B: Determining the Required Pendulum Length for a Given Beats Per Hour Rate

BPM = Beat Per Minute   BPH = Beats Per Hour
P LEN = Pendulum Length in Inches (to center of Bob, assumes pendulum stick/rod has no weight)

BPM	BPH	P LEN	BPM	BPH	P LEN	BPM	BPH	P LEN	BPM	BPH	P LEN
60	3600	39.1	101	6060	13.8	142	8520	7.0	183	10980	4.2
61	3660	37.8	102	6120	13.5	143	8580	6.9	184	11040	4.2
62	3720	36.6	103	6180	13.3	144	8640	6.8	185	11100	4.1
63	3780	35.5	104	6240	13.0	145	8700	6.7	186	11160	4.1
64	3840	34.4	105	6300	12.8	146	8760	6.6	187	11220	4.0
65	3900	33.3	106	6360	12.5	147	8820	6.5	188	11280	4.0
66	3960	32.3	107	6420	12.3	148	8880	6.4	189	11340	3.9
67	4020	31.4	108	6480	12.1	149	8940	6.3	190	11400	3.9
68	4080	30.4	109	6540	11.8	150	9000	6.3	191	11460	3.9
69	4140	29.6	110	6600	11.6	151	9060	6.2	192	11520	3.8
70	4200	28.7	111	6660	11.4	152	9120	6.1	193	11580	3.8
71	4260	27.9	112	6720	11.2	153	9180	6.0	194	11640	3.7
72	4320	27.2	113	6780	11.0	154	9240	5.9	195	11700	3.7
73	4380	26.4	114	6840	10.8	155	9300	5.9	196	11760	3.7
74	4440	25.7	115	6900	10.6	156	9360	5.8	197	11820	3.6
75	4500	25.0	116	6960	10.5	157	9420	5.7	198	11880	3.6
76	4560	24.4	117	7020	10.3	158	9480	5.6	199	11940	3.6
77	4620	23.7	118	7080	10.1	159	9540	5.6	200	12000	3.5
78	4680	23.1	119	7140	9.9	160	9600	5.5	201	12060	3.5
79	4740	22.6	120	7200	9.8	161	9660	5.4	202	12120	3.4
80	4800	22.0	121	7260	9.6	162	9720	5.4	203	12180	3.4
81	4860	21.5	122	7320	9.5	163	9780	5.3	204	12240	3.4
82	4920	20.9	123	7380	9.3	164	9840	5.2	205	12300	3.3
83	4980	20.4	124	7440	9.2	165	9900	5.2	206	12360	3.3
84	5040	19.9	125	7500	9.0	166	9960	5.1	207	12420	3.3
85	5100	19.5	126	7560	8.9	167	10020	5.0	208	12480	3.3
86	5160	19.0	127	7620	8.7	168	10080	5.0	209	12540	3.2
87	5220	18.6	128	7680	8.6	169	10140	4.9	210	12600	3.2
88	5280	18.2	129	7740	8.5	170	10200	4.9	211	12660	3.2
89	5340	17.8	130	7800	8.3	171	10260	4.8	212	12720	3.1
90	5400	17.4	131	7860	8.2	172	10320	4.8	213	12780	3.1
91	5460	17.0	132	7920	8.1	173	10380	4.7	214	12840	3.1
92	5520	16.6	133	7980	8.0	174	10440	4.6	215	12900	3.0
93	5580	16.3	134	8040	7.8	175	10500	4.6	216	12960	3.0
94	5640	15.9	135	8100	7.7	176	10560	4.5	217	13020	3.0
95	5700	15.6	136	8160	7.6	177	10620	4.5	218	13080	3.0
96	5760	15.3	137	8220	7.5	178	10680	4.4	219	13140	2.9
97	5820	15.0	138	8280	7.4	179	10740	4.4	220	13200	2.9
98	5880	14.7	139	8340	7.3	180	10800	4.3	221	13260	2.9
99	5940	14.4	140	8400	7.2	181	10860	4.3	222	13320	2.9
100	6000	14.1	141	8460	7.1	182	10920	4.2	223	13380	2.8

Chart C: Spring Strengths of Old Versus New Springs

The strength of a spring varies as the cube of its thickness. Therefore, if you have an old spring of one thickness and wish to use a new spring of a different thickness, this chart will give you an estimate of the relative strength of the new spring versus the old one.

- Old spring thickness reads down the left side.
- New spring thickness reads across the top of the chart.
- Number in the chart is the relative difference (as a multiplier) between spring strengths.

Examples:

A new spring 15 mils thick is 2 times as strong as an old spring 12 mils thick.

A new spring 16 mils thick is only .70 times as strong as an old spring 18 mils thick.

Old Spring Thickness (Mils) — New Spring Thickness In Mils (In Thousandths Of An Inch)

Old	8	9	10	11	12	13	14	15	16	17	18	19	20	21	22	23	24	25	26	27	28
8	1.0	1.4	2.0	2.6	3.4	4.3	5.4	6.6	8.0	9.6	11.4	13.4	15.6	18.1	20.8	23.8	27.0	30.5	34.3	38.4	42.9
9	0.7	1.0	1.4	1.8	2.4	3.0	3.8	4.6	5.6	6.7	8.0	9.4	11.0	12.7	14.6	16.7	19.0	21.4	24.1	27.0	30.1
10	0.5	0.7	1.0	1.3	1.7	2.2	2.7	3.4	4.1	4.9	5.8	6.9	8.0	9.3	10.6	12.2	13.8	15.6	17.6	19.7	22.0
11	0.4	0.5	0.8	1.0	1.3	1.7	2.1	2.5	3.1	3.7	4.4	5.2	6.0	7.0	8.0	9.1	10.4	11.7	13.2	14.8	16.5
12	0.3	0.4	0.6	0.8	1.0	1.3	1.6	2.0	2.4	2.8	3.4	4.0	4.6	5.4	6.2	7.0	8.0	9.0	10.2	11.4	12.7
13	0.2	0.3	0.5	0.6	0.8	1.0	1.2	1.5	1.9	2.2	2.7	3.1	3.6	4.2	4.8	5.5	6.3	7.1	8.0	9.0	10.0
14	0.2	0.3	0.4	0.5	0.6	0.8	1.0	1.2	1.5	1.8	2.1	2.5	2.9	3.4	3.9	4.4	5.0	5.7	6.4	7.2	8.0
15	0.2	0.2	0.3	0.4	0.5	0.7	0.8	1.0	1.2	1.5	1.7	2.0	2.4	2.7	3.2	3.6	4.1	4.6	5.2	5.8	6.5
16	-	0.2	0.2	0.3	0.4	0.5	0.7	0.8	1.0	1.2	1.4	1.7	2.0	2.3	2.6	3.0	3.4	3.8	4.3	4.8	5.4
17	-	-	0.2	0.3	0.4	0.4	0.6	0.7	0.8	1.0	1.2	1.4	1.6	1.9	2.2	2.5	2.8	3.2	3.6	4.0	4.5
18	-	-	0.2	0.2	0.3	0.4	0.5	0.6	0.7	0.8	1.0	1.2	1.4	1.6	1.8	2.1	2.4	2.7	3.0	3.4	3.8
19	-	-	-	0.2	0.3	0.3	0.4	0.5	0.6	0.7	0.9	1.0	1.2	1.4	1.6	1.8	2.0	2.3	2.6	2.9	3.2
20	-	-	-	0.2	0.2	0.3	0.3	0.4	0.5	0.6	0.7	0.9	1.0	1.2	1.3	1.5	1.7	2.0	2.2	2.5	2.7
21	-	-	-	-	0.2	0.2	0.3	0.4	0.4	0.5	0.6	0.7	0.9	1.0	1.1	1.3	1.5	1.7	1.9	2.1	2.4
22	-	-	-	-	0.2	0.2	0.3	0.3	0.4	0.5	0.5	0.6	0.8	0.9	1.0	1.1	1.3	1.5	1.7	1.8	2.1
23	-	-	-	-	-	0.2	0.2	0.3	0.3	0.4	0.5	0.6	0.7	0.8	0.9	1.0	1.1	1.3	1.4	1.6	1.8
24	-	-	-	-	-	0.2	0.2	0.2	0.3	0.4	0.4	0.5	0.6	0.7	0.8	0.9	1.0	1.1	1.3	1.4	1.6
25	-	-	-	-	-	-	0.2	0.2	0.3	0.3	0.4	0.4	0.5	0.6	0.7	0.8	0.9	1.0	1.1	1.3	1.4
26	-	-	-	-	-	-	0.2	0.2	0.2	0.3	0.3	0.4	0.5	0.5	0.6	0.7	0.8	0.9	1.0	1.1	1.2
27	-	-	-	-	-	-	-	0.2	0.2	0.2	0.3	0.3	0.4	0.5	0.5	0.6	0.7	0.8	0.9	1.0	1.1
28	-	-	-	-	-	-	-	0.2	0.2	0.2	0.3	0.3	0.4	0.4	0.5	0.6	0.6	0.7	0.8	0.9	1.0

Spring strength is a direct function of spring width. One spring 10% wider than another will be 10% stronger, all other factors being equal.

To estimate relative strength using both width and thickness, use the following calculation:

Relative strengths between 2 springs =

(New spring width/old spring width) X (# for springs found in chart above)

Example:

New spring 0.703 wide and 0.16 thick, old spring 0.750 wide and 0.10 thick

(0.703/0.750)(1.7) = 1.59 times relative strength

Understand that the chart shows estimated values. A spring made in the year 1725 will not be the same metallurgy as a spring made in the year 2005. Even if the springs have the same dimensions, springs from different manufacturers may vary in strength.

Franz Hermle Company

When a clock comes across your work bench that needs major work, the first thing you have to do is decide whether to rebuild the clock movement, or to replace it. In making this decision, you have to compare the price of a replacement movement with the labor cost of rebuilding the existing one (which may amount to quite a bit, given that rebuilding means total disassembly/reassembly).

One of the largest (if not THE largest) manufacturers of clock movements today is the Franz Hermle Company. They are an excellent source of replacement parts and clock movements, and you'll find that their spring and barrel combinations are well documented in most of the major supplier catalogs.

You may have heard some negative comments about Hermle, due mostly to events back in the 1970's (when, even though they were not the only clock manufacturer back then to use poor metals, they got a bad rap for doing so). My experience with their products, however, has been extremely positive: as of the writing of this book, I have replaced about 70 old Hermle movements with new Hermle movements, with absolutely no problems or complaints about any of the new ones. The average age of the older ones I have replaced has been about 30 years, which is a reasonable lifespan for most clock movements. They are one of the rare mass production companies today that have been in business for many years, yet are still making product improvements on a regular basis. Some of the improvements have obviously raised the cost of their products over the years, but perhaps the improvements are why they have been so successful for so long.

Some of the advantages to using Hermle movements and parts:

- They make drop-in replacement movements. You can order a brand new replacement for most any movement made in the last 30 years or so, and have it within 48 hours.

- All of the function synchronization on Hermle movements is done outside the plates. This allows a repairman to just take the movement inside the plates apart, clean it and put the parts inside the plates back together. They can then do all the synchronization adjustments afterward.

  This means that you don't have to take the plates apart if any wheel mesh is off by one or two teeth.

  This also means that you can drop out the barreled springs as an assembly without taking the plates apart.

- Each barrel is stamped with a number which totally identifies it and its contained spring, making it easy to determine replacements.

- Their movements are stamped with a date code to help assess whether or not to replace an old movement (i.e.: makes it easy to determine how old a movement is).

- Various specifications for many Hermle barreled springs can be found in most major supplier catalogs (usually listed by the Hermle barrel number).

This Page Intentionally Blank.

Chart D: American Clock Springs by Clock Model

Size (In Inches)	End	Description

Ansonia

Size (In Inches)	End	Description
5/32 x .010 x 30	Hole	For 1W
5/16 x .015 x 42	Loop	For E167 Time
3/8 x .009 x 39-3/4	Loop	For Quick Tick Time
1/2 x .016 x 66	Hole	
3/8 x .014 x 60	Loop	Time spring for Amazone, Clatter, Flash, Pirate, Presto, Princess, Racket, Rally, Repeater, Startle, Tattler, Trolley and Trump
3/8 x .011 x 51	Loop	Strike spring for Carriage and Racket. Time spring for Tourist
1/4 x .010 x 26	Loop	Alarm spring for Amazone, Flash, Pirate, Presto, Racket, Rally and Tourist. Time spring for Carriage
3/8 x .013 x 34	Loop	Alarm spring for Clatter, Tattler and Trump
1/4 x .012 x 41	Loop	Time spring for Midget
3/8 x .017 x 54	Hole	Time spring for Peep O Day
7/16 x .014 x 72	Loop	Alarm spring for Repeater and Trolley
5/8 x .013 x 69	Loop	Time spring for Simplex
1/4 x .013 x 24	Loop	Time spring for Spark and Tot. Alarm spring for Tot
5/8 x .013 x 105	Hole	Time spring for Spark and 8 Day
3/8 x .017 x 96	Loop	For 8-Day Capitol Clock
9/16 x .018 x 96	Loop	For 8-Day Lever and Bungalo Clock
3/4 x .012 x 90	Hole	For Marquise Time and Strike

Chelsea Clock Company

Size (In Inches)	End	Description
17/32 x .010 x 72	Hole	Time spring
43/64 x .010 x 72	Hole	Strike spring
1/2 x .012 x 39	Hole	Auto clock

China Plate

Size (In Inches)	End	Description
1/2 x .015 x 36	Hole	
7/16 x .018 x 48	Loop	

Gilbert

Size (In Inches)	End	Description
3/8 x .016 x 60	Loop	Time spring for All Time, Bi-Nite, Connecticut, Design, Radium, Tornado, Wake-up Nos. 4582 to 4587
5/16 x .009 x 36	Loop	Alarm spring for All Time, Connecticut, Design, Radium Nos. 4582 to 4585 and Nos. 4601 and 4602
5/16 x .009 x 20	Loop	Alarm spring for Bi-Nite, Tornado, Wake-up Nos. 4570, 4581, 4582, 4584, 4586 and 4587
9/16 x .016 x 66	Loop	Time spring for "Nine" Clock and 8-Day Lever Clock
1/2 x .016 x 66	Loop	Alarm spring for "Nine" Clock
3/4 x .017 x 120	Loop	For 8-Day Pendulum, Time and Strike
1/2 x .013 x 60	Hole	Time Spring for 4000 Series
1/2 x .008 x 78	Hole	Strike spring for 4000 Series
5/16 x .017 x 42	Loop	1-Day Alarm
3/4 x .018 x 96	Loop	For 8-Day Pendulum, Time and Strike

Chart D: American Clock Springs by Clock Model (Continued)

==

Size (In Inches)	End	Description

==

Ingraham

Size (In Inches)	End	Description
1/2 x .016 x 66	Loop	Time spring for Ace and Victory, Miller Plate, Miller Kitchen and Victorian, National Call
13/32 x .014 x 54	Loop	Alarm spring for Ace and Victory
17/32 x .015 x 40	Loop	Time and Alarm spring for O.M. Admiral
3/8 x .016 x 60	Loop	Time spring for N.M. Admiral, Autocrat, Liberty, Comet, Dawn, Flyer, Indian, Rattler, Sentinel, Sentry, "Style" Queen, Vim, X-ray, Liberator, Liberty
9/32 x .017 x 48	Loop	Alarm spring for N.M. Admiral, Autocrat, Liberty and Victory
1/4 x .010 x 26	Loop	Alarm spring for Comet, Dawn, Flyer, Indian, Rattler, Sentinel, Sentry and X-ray
3/4 x .018 x 96	Loop	Time Strike spring for 8-Day Parlor and Mantle and Office Clock
1/4 x .011 x 33-1/2	Hole	Little Pal

New Haven

Size (In Inches)	End	Description
1/2 x .015 x 84	Loop	Time spring for Automatic Signal, Ti-Tan, 8-Day. Alarm spring for Automatic Signal, Ti-Tan, 8-Day
3/8 x .016 x 60	Loop	Time spring for Beacon, Commuter, N.S., Tell Tale and Tick Tock, Prompter, Abbey, Artlarm, St. Elmo, Recall and Tom Tom
1/4 x .010 x 26	Loop	Alarm spring for Beacon
9/32 x .009 x 30	Loop	Time spring for Brownie, Tattoo Junior, Artlarm, O.S. Tell Tale, Tick Tock and Tidy Tot
7/32 x .011 x 24	Loop	Alarm spring for Brownie, O.M. Tattoo Junior
3/8 x .013 x 34	Loop	Alarm spring for Commuter, Prompter, Slumber Stopper, Tattoo and Tom Tom
5/32 x .009 x 26	Loop	Alarm spring for N.M. Tatoo Junior, Tell Tale, Tick Tock and Tiny Tot
5/16 x .010 x 42	Hole	Alarm spring for Recall
9/16 x .015 x 78	Loop	Time spring for Petite Time-Tambour
3/4 x .018 x 96	Loop	Time spring for 8-Day Combo
3/4 x .017 x 120	Loop	Time and Strike spring for 8-Day Combo No. 2
3/4 x .012 x 72	Loop	Time and Strike spring for Class Reg
7/8 x .013 x 72	Hole	Time and Strike spring for Chime Clock
3/4 x .009 x 72	Hole	Strike spring for Chime Clock
3/4 x .014 x 90	Hole	Time spring for Harmonious and Duo Clock and No. 180
3/4 x .012 x 90	Hole	Strike spring for Harmonious and Duo Clock and No. 180
5/8 x .014 x 70	Hole	For Ormond 8-Day Clock

Parker Clock Company

Size (In Inches)	End	Description
1/4 x .011 x 39	Hole	Alarm spring
3/8 x .013 x 34	Loop	Alarm spring
3/8 x .014 x 60	Loop	Alarm spring

Phinney Walker Clock Company

Size (In Inches)	End	Description
5/8 x .016 x 59	Hole	
3/4 x .013 x 54	Hole	
3/16 x .009 x 24	Hole	

Chart D: American Clock Springs by Clock Model (Continued)

===

Size (In Inches)	End	Description

===

Sessions

Size (In Inches)	End	Description
5/8 x .016 x 59	Hole	Time and Alarm spring for all 8-Day Alarm Clocks
3/8 x .017 x 96	Loop	Time spring for Verdi Clock
11/16 x .018 x 96	Loop	Stike spring for Verdi Clock
1/4 x .013 x 24	Loop	For Kitchen Alarm, Jigger
5/16 x .010 x 28	Hole	For 30 Hour Clocks
5/8 x .014 x 70	Hole	For 8-Day Marine and Ormond Clocks
3/4 x .018 x 96	Loop	For 8-Day Parlor, Kitchen and Banjo Clocks. Time, Strike and Chime spring for Chime Clocks
5/8 x .013 x 69	Loop	For all late Model Alarm Clocks, Time and Alarm

Seth Thomas

Size (In Inches)	End	Description
3/8 x .016 x 60	Loop	Movement Nos. 1, 3 and 109. Time spring
5/8 x .018 x 72	Hole	Movement No. 2 Time spring. Ships Bell
1/2 x .018 x 96	Loop	Movement No. 3. Alarm spring
5/16 x .016 x 54	Loop	Movement No. 4. Strike spring
7/16 x .018 x 60	Loop	Movement No. 6. Time Spring
5/8 x .013 x 78	Hole	Movement Nos. 7 and 103A. Time spring
3/8 x .024 x 108	Hole	Movement No. 10A. Time spring
5/8 x .018 x 96	Hole	Movement No. 11. Strike spring
3/4 x .012 x 90	Hole	Movement Nos. 22 and 48. Time spring
11/16 x .015 x 108	Loop	Movement Nos. 41 and 44. Time spring
3/4 x .016 x 60	Hole	Movement Nos. 48 and 117. Strike spring
3/4 x 0.13 x 54	Hole	Movement No. 51. Time and Strike spring Movement No. 105. Strike spring
3/4 x .017 x 120	Loop	Movement Nos. 89 and 123. Time and Strike spring. Time and Strike. York Nos. 1-2-5-6-7 Tambour
5/16 x .010 x 32	Hole	Movement No. 102. Time Petite Clock
1/2 x .016 x 65	Hole	Movement No. 103. Time spring
3/4 x .016 x 60	Hole	Movement No. 105. Time spring
3/16 x .013 x 24	Loop	Movement No. 109. Alarm spring
1/2 x .016 x 66	Loop	Movement No. 110. Time and Alarm spring
9/16 x .012 x 39	Hole	Movement Nos 112 and 178
7/8 x .018 x 84	Hole	Movement No. 113. Time and Strike spring
1 x .018 x 96	Hole	Movement No. 113AB. Time and Strike spring and Chime spring
3/4 x .011 x 50	Hole	Movement No. 115A. Strike spring
11/16 x .014 x 54	Hole	Movement No. 115B. Time spring Movement No. 120. Time and Strike spring Movement No. 124. Time and Strike spring Movement No. 125. Strike spring
1/2 x .013 x 60	Hole	Movement No.116. Time spring
3/4 x .015 x 72	Hole	Movement No. 124 Chime spring. Whitby Mahog. Cab.
11/16 x .012 x 72	Hole	Movement No. 125. Time spring
5/16 x .016 x 72	Loop	Movement No. 673
3/8 x .011 x 48	Hole	Movement No. 5100

Chart D: American Clock Springs by Clock Model (Continued)

===

Size (In Inches)	End	Description

===

Waterbury

Size (In Inches)	End	Description
3/8 x .014 x 48	Loop	Time spring for Must-Get-Up, Spasmodic, Sunrise, Turncut, Twin, Vanatie, Vigilant. 1-Day Lever
5/16 x .008 x 27	Hole	Time for Rattler, Tourist, Wanderer and Wasp
1/2 x .018 x 60	Loop	Alarm spring for Must-Get-Up, Relay and Spasmodic
3/16 x .013 x 24	Loop	Alarm spring for Rattler, Tourist, Wasp and Relay Junior
3/8 x .014 x 48	Hole	Time spring for Relay, Relay Juniour, Royal Thrift
7/32 x .011 x 24	Loop	Alarm spring for Royal, Sunrise, Thrift, Turn-Out, Twin, Vigilant and No-Sleep
5/32 x .008 x 24	Hole	Alarm spring for Skip and Wanderer and 1-Day Lever Strike
5/16 x .013 x 24	Loop	Alarm spring for Vanitie
5/16 x .016 x 54	Hole	Spring for 1-Day Lever
3/4 x .012 x 75	Hole	Strike spring for Mantel Clock. Time spring for 1-Day Cabinet
3/4 x .012 x 72	Loop	Strike spring for Mantel Clock
3/4 x .016 x 78	Loop	Strike spring for Mantel Clock
3/4 x .017 x 120	Loop	Time spring for Chime 34, 34D, 35 and 30 Movement, also for 337 to 340 Movements
3/4 x .014 x 108	Loop	30-Day Clock Time Side (has no strike). Strike spring for Chimes 34, 34D, 30 and 35. Also for 337 to 340 Movements
1 x .018 x 96	Hole	Time for 8-Day Lever Clock
3/8 x .015 x 53	Loop	Mickey Mouse
5/16 x .012 x 54	Riveted	Johour Alarm (U.S. Time) Waterbury #30-50033 Special

Below: two springs, good (new) and bad (used)

Facing page: comparison of the 2 springs with more context

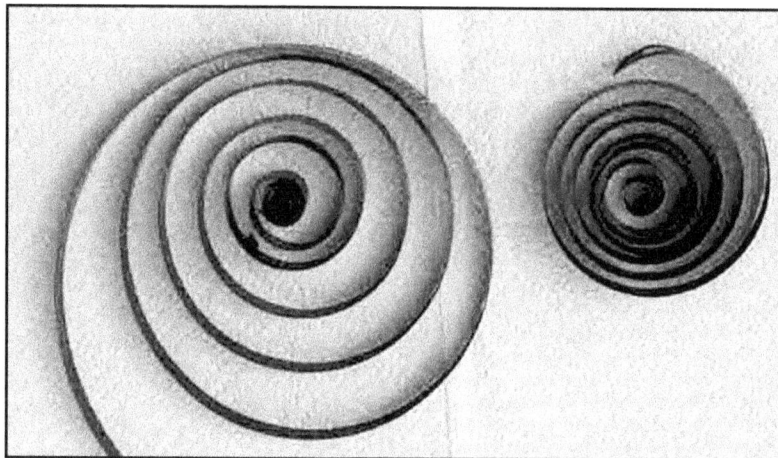

Good New Spring   Bad Used Spring

This Page Intentionally Blank.

About The Author

Richard Hansen is an electrical engineer who graduated from Auburn University. He worked for years in California in the Silicon Valley as a design engineer. After taking night classes in clock repair, the hobby interest in clocks he had developed over the years grew to become a career change.

Richard is a second generation clock person (his father owned 800 clocks at one time). Since changing careers, Richard has taught beginning and advanced clock repair in California, Pennsylvania, New York, and Connecticut - to both amateurs and professional clock repairmen. For the last fifteen years, he has owned and operated a clock repair store in New York state.

Color Prints of Model Photos in Book Available

If you would like to have a *free* color 4 x 6 print of any of the women model photos in this book (on the cover or on page 64), send your request to:

richardhansen6969@gmail.com

Also By Richard Hansen:

Available from Amazon.com and dozens of other on-line book retailers

(Available at wholesale pricing to qualified retailers - email: questions@goofyrooster-publishing.com)

Suggested Retail Price $7.95

Some clock repairmen give away a copy of this book to their customers with every repair job and/or clock purchase!

How to Fix Your Own Clock gives simple answers to basic minor problems that can be easily corrected by most clock owners. This book is written for the typical mechanical clock owner who knows little about clocks - and who doesn't want to run up a repair bill if it can be helped. The book is in an easy to understand question and answer format, and comes from actual questions from the author's newspaper column.

The author, Richard Hansen (a clock repairman and owner of his own clock repair business for more than 15 years), knows that often, a clock may need little more than the correct set-up, or a minor adjustment to go from "dust collector" in your attic - to "treasured and accurate timepiece" in your living room! He wrote this book so you can get your clock going - and keep it going - easily and inexpensively!